D1013010

GENETICS, CELL DIFFERENTIATION, AND CANCER

BRISTOL-MYERS CANCER SYMPOSIA

Series Editor
MAXWELL GORDON
Science and Technology Group
Bristol-Myers Company

1. Harris Busch, Stanley T. Crooke, and Yerach Daskal (Editors).
Effects of Drugs on the Cell Nucleus, 1979.

2. Alan C. Sartorelli, John S. Lazo, and Joseph R. Bertino (Editors).
Molecular Actions and Targets for Cancer Chemotherapeutic Agents, 1981.

3. Saul A. Rosenberg and Henry S. Kaplan (Editors).
Malignant Lymphomas: Etiology, Immunology, Pathology, Treatment, 1982.

4. Albert H. Owens, Jr., Donald S. Coffey, and Stephen B. Baylin (Editors).
Tumor Cell Heterogeneity: Origins and Implications, 1982.

5. Janet D. Rowley and John E. Ultmann (Editors).
Chromosomes and Cancer: From Molecules to Man, 1983.

6. Umberto Veronesi and Gianni Bonadonna (Editors).
Clinical Trials in Cancer Medicine: Past Achievements and Future Prospects, 1985.

7. Paul A. Marks (Editor).
Genetics, Cell Differentiation, and Cancer, 1985.

GENETICS, CELL DIFFERENTIATION, AND CANCER

Edited by

PAUL A. MARKS
DeWitt Wallace Research Laboratory
Sloan-Kettering Institute
Memorial Sloan-Kettering Cancer Center
New York, New York

1985
ACADEMIC PRESS, INC.
Harcourt Brace Jovanovich, Publishers
Orlando San Diego New York Austin
London Montreal Sydney Tokyo Toronto

ACADEMIC PRESS, INC.
Orlando, Florida 32887

United Kingdom Edition published by
ACADEMIC PRESS INC. (LONDON) LTD.
24–28 Oval Road, London NW1 7DX

LIBRARY OF CONGRESS CATALOGING IN PUBLICATION DATA

Main entry under title:

Genetics, cell differentiation, and cancer.

(Bristol-Myers cancer symposia ; 7)

Proceedings of the Seventh Annual Symposium on Cancer Research, held at the Memorial Sloan-Kettering Cancer Center, in Oct. 1984, on the occasion of its centennial year celebration in 1984, sponsored by the Bristol-Myers Company.

Includes index.

1. Cancer—Genetic aspects—Congresses. 2. Cell transformation—Congresses. 3. Cell differentiation—Congresses. 4. Oncogenes—Congresses. 5. Gene expression—Congresses. I. Marks, Paul A. II. Bristol-Myers Symposium on Cancer Research (7th : 1984 : Memorial Sloan-Kettering Cancer Center) III. Memorial Sloan-Kettering Cancer Center. IV. Bristol-Myers Company. V. Series. [DNLM: 1. Cell Differentiation—congresses. 2. Gene Expression Regulation—congresses. 3. Oncogenes—congresses. 4. Oncogenic Viruses—genetics—congresses. 5. Neoplasms—familial & genetic—congresses. W3 BR429 v.7 / QZ 202 G3317 1984]

RC268.4.G455 1985 616.99'4071 85-9027

ISBN 0-12-473060-4 (alk. paper)
PRINTED IN THE UNITED STATES OF AMERICA

85 86 87 88 9 8 7 6 5 4 3 2 1

Contents

10 Differences in the Biological Function of Viral and Cellular *src* Genes

HIDEO IBA, TATSUO TAKEYA, FREDERICK R. CROSS, ELLEN A. GARBER, TERUKO HANAFUSA, DAVID PELLMAN, and HIDESABURO HANAFUSA

11 Function of Yeast *RAS* Genes

MICHAEL WIGLER, SCOTT POWERS, TOHRU KATAOKA, TAKASHI TODA, OTTAVIO FASANO, KUNIHIRO MATSUMOTO, ISAO UNO, TATSUO ISHIKAWA, and JAMES BROACH

12 The *neu* Oncogene Encodes a Cell Surface Protein with Properties of a Growth Factor Receptor

DAVID F. STERN, ALAN SCHECHTER, LALITHA VAIDYANATHAN, MARK GREENE, JEFFREY DREBIN, and ROBERT WEINBERG

13 Cooperativity between "Primary" and "Auxiliary" Oncogenes of Defective Avian Leukemia Viruses

THOMAS GRAF, BECKY ADKINS, ACHIM LEUTZ, HARTMUT BEUG, and PATRICIA KAHN

The Family of Human T-Lymphotropic Retroviruses Called Human T-Cell Leukemia/Lymphoma Virus (HTLV): Their Role in Lymphoid Malignancies and Lymphosuppressive Disorders (AIDS)

R. C. GALLO, L. RATNER, M. POPOVIC, S. Z. SALAHUDDIN, M. G. SARNGADHARAN, F. WONG-STAAL, G. SHAW, B. HAHN, P. D. MARKHAM, J. GROOPMAN, B. SAFAI, M. REITZ, and M. ROBERT-GUROFF

Contributors

Numbers in parentheses indicate the pages on which the authors' contributions begin.

SAMIT ADHYA (15), Department of Molecular Biology and Virology, Memorial Sloan-Kettering Institute for Cancer Research, New York, New York 10021

BECKY ADKINS[1] (171), European Molecular Biology Laboratory, 6900 Heidelberg, Federal Republic of Germany

KAREN ARTZT (95), Memorial Sloan-Kettering Cancer Center, New York, New York 10021

DOROTHEA BENNETT (95), Memorial Sloan-Kettering Cancer Center, New York, New York 10021

HARTMUT BEUG (171), European Molecular Biology Laboratory, 6900 Heidelberg, Federal Republic of Germany

J. MICHAEL BISHOP (135), Department of Microbiology and Immunology, and The George Williams Hooper Foundation, University of California Medical Center, San Francisco, California 94143

J.-F. BRIAT (47), Department of Biochemistry, University of California, Berkeley, California 94720

JAMES BROACH (153), Department of Molecular Biology, Princeton University, Princeton, New Jersey 08540

MICHAEL J. CHAMBERLIN (47), Department of Biochemistry, University of California, Berkeley, California 94720

D. F. CLAYTON (119), The Rockefeller University, New York, New York 10021

[1]Present address: Department of Pathology, Stanford University Medical Center, Stanford, California 94305.

B. CLURMAN (127), Molecular Biology and Virology Program of the Graduate School, Memorial Sloan-Kettering Cancer Center, New York, New York 10021

FREDERICK R. CROSS (143), The Rockefeller University, New York, New York 10021

J. E. DARNELL, JR. (119), The Rockefeller University, New York, New York 10021

RUSSELL L. DEDRICK (47), Department of Biochemistry, University of California, Berkeley, California 94720

JEFFREY DREBIN (165), Department of Rheumatology, Tufts University School of Medicine, Boston, Massachusetts

OTTAVIO FASANO (153), Cold Spring Harbor Laboratory, Cold Spring Harbor, New York 11724

JEFFREY FIELD (15), Department of Molecular Biology and Virology, Memorial Sloan-Kettering Institute for Cancer Research, New York, New York 10021

J. M. FRIEDMAN (119), The Rockefeller University, New York, New York 10021

R. C. GALLO (183), Laboratory of Tumor Cell Biology, National Cancer Institute, Bethesda, Maryland 20205

ELLEN A. GARBER (143), The Rockefeller University, New York, New York 10021

M. GOODENOW (127), Molecular Biology and Virology, Program of the Graduate School, Memorial Sloan-Kettering Cancer Center, New York, New York 10021

THOMAS GRAF (171), European Molecular Biology Laboratory, 6900 Heidelberg, Federal Republic of Germany

MARK GREENE (165), Department of Rheumatology, Tufts University School of Medicine, Boston, Massachusetts

RICHARD GRONOSTAJSKI[2] (15), Department of Molecular Biology and Virology, Memorial Sloan-Kettering Institute for Cancer Research, New York, New York 10021

J. GROOPMAN (183), Hematology/Oncology Division, Deaconess Medicine, Boston, Massachusetts 02215

RONALD A. GUGGENHEIMER (15), Department of Molecular Biology and Virology, Memorial Sloan-Kettering Institute for Cancer Research, New York, New York 10021

[2]Present address: Department of Medical Genetics, University of Toronto, Toronto, Ontario, Canada M55 1A8.

B. HAHN (183), Laboratory of Tumor Cell Biology, National Cancer Institute, Bethesda, Maryland 20205

HIDESABURO HANAFUSA (143), The Rockefeller University, New York, New York 10021

TERUKO HANAFUSA (143), The Rockefeller University, New York, New York 10021

MICHELLE HANNA (47), Department of Biochemistry, University of California, Berkeley, California 94720

A. HAYDAY (127), Massachusetts Institute of Technology, Cambridge, Massachusetts 12139

W. HAYWARD (127), Molecular Biology and Virology Program of the Graduate School, Memorial Sloan-Kettering Cancer Center, New York, New York 10021

JERARD HURWITZ (15), Department of Molecular Biology and Virology, Memorial Sloan-Kettering Institute for Cancer Research, New York, New York 10021

HIDEO IBA[3] (143), The Rockefeller University, New York, New York 10021

TATSUO ISHIKAWA (153), Institute of Applied Microbiology, University of Tokyo, Toyko 113, Japan

PATRICIA KAHN (171), European Molecular Biology Laboratory, 6900 Heidelberg, Federal Republic of Germany

CAROLINE M. KANE (47), Department of Biochemistry, University of California, Berkeley, California 94720

TOHRU KATAOKA (153), Cold Spring Harbor Laboratory, Cold Spring Harbor, New York 11724

MARK KENNY (15), Department of Molecular Biology and Virology, Memorial Sloan-Kettering Institute for Cancer Research, New York, New York 10021

R. LESTRANGE (127), Molecular Biology and Virology, Program of the Graduate School, Memorial Sloan-Kettering Cancer Center, New York, New York 10021

ACHIM LEUTZ (171), European Molecular Biology Laboratory, 6900 Heidelberg, Federal Republic of Germany

JUDITH LEVIN (47), Department of Biochemistry, University of California, Berkeley, California 94720

[3]Present address: Department of Biophysics and Biochemistry, Faculty of Science, The University of Tokyo, Tokyo 113, Japan.

JEFF LINDENBAUM (15), Department of Molecular Biology and Virology, Memorial Sloan-Kettering Institute for Cancer Research, New York, New York 10021

P. D. MARKHAM (183), Department of Cell Biology, Litton Bionetics, Inc., Kensington, Maryland 20895

PAUL A. MARKS (1, 105), DeWitt Wallace Research Laboratory, Sloan-Kettering Institute, Memorial Sloan-Kettering Cancer Center, New York, New York 10021

KUNIHIRO MATSUMOTO[4] (153), Department of Industrial Chemistry, Tottori University, Tottori 680, Japan

KYOUSUKE NAGATA[5] (15), Department of Molecular Biology and Virology, Memorial Sloan-Kettering Institute for Cancer Research, New York, New York 10021

HOWARD A. NASH (75), Laboratory of Molecular Biology, National Institute of Mental Health, Bethesda, Maryland 20205

DAVID PELLMAN (143), The Rockefeller University, New York, New York 10021

M. POPOVIC (183), Laboratory of Tumor Cell Biology, National Cancer Institute, Bethesda, Maryland 20205

D. J. POWELL (119), The Rockefeller University, New York, New York 10021

SCOTT POWERS (153), Cold Spring Harbor Laboratory, Cold Spring Harbor, New York 11724

L. RATNER[6] (183), Laboratory of Tumor Cell Biology, National Cancer Institute, Bethesda, Maryland 20205

M. REITZ (183), Laboratory of Tumor Cell Biology, National Cancer Institute, Bethesda, Maryland 20205

REBECCA REYNOLDS (47), Department of Biochemistry, University of California, Berkeley, California 94720

ALEXANDER RICH (25), Department of Biology, Massachusetts Institute of Technology, Cambridge, Massachusetts 02139

R. A. RIFKIND (105), DeWitt Wallace Research Laboratory, Sloan-Kettering Institute, Memorial Sloan-Kettering Cancer Center, New York, New York 10021

[4]Present address: DNAX Research Institute of Molecular and Cellular Biology, Palo Alto, California 94304.

[5]Present address: Department of Molecular Genetics, National Institute of Genetics, Mishima, Shizuoka, Japan.

[6]Present address: Division of Hematology and Oncology, Washington University, St. Louis, Missouri.

M. ROBERT-GUROFF (183), Laboratory of Tumor Cell Biology, National Cancer Institute, Bethesda, Maryland 20205

B. SAFAI (183), Memorial Sloan-Kettering Institute, New York, New York 10021

S. Z. SALAHUDDIN (183), Laboratory of Tumor Cell Biology, National Cancer Institute, Bethesda, Maryland 20205

M. G. SARNGADHARAN (183), Laboratory of Tumor Cell Biology, National Cancer Institute, Bethesda, Maryland 20205

ALAN SCHECHTER (165), Whitehead Institute for Biomedical Research, Massachusetts Institute of Technology, Cambridge, Massachusetts 02142

MARTIN SCHMIDT (47), Department of Biochemistry, University of California, Berkeley, California 94720

G. SHAW[7] (183), Laboratory of Tumor Cell Biology, National Cancer Institute, Bethesda, Maryland 20205

M. SHEFFERY (105), DeWitt Wallace Research Laboratory, Sloan-Kettering Institute, Memorial Sloan-Kettering Cancer Center, New York, New York 10021

C.-K. SHIH (127), Molecular Biology and Virology Program of the Graduate School, Memorial Sloan-Kettering Cancer Center, New York, New York 10021

HEE-SUP SHIN (95), Memorial Sloan-Kettering Cancer Center, New York, New York 10021

M. SIMON (127), Molecular Biology and Virology, Program of the Graduate School, Memorial Sloan-Kettering Cancer Center, and The Rockefeller University, New York, New York 10021

DAVID F. STERN (165), Whitehead Institute for Biomedical Research, Massachusetts Institute of Technology, Cambridge, Massachusetts 02142

TATSUO TAKEYA[8] (143), The Rockefeller University, New York, New York 10021

TAKASHI TODA (153), Cold Spring Harbor Laboratory, Cold Spring Harbor, New York 11724

S. TONEGAWA (127), Massachusetts Institute of Technology, Cambridge, Massachusetts 12139

[7]Present address: Division of Hematology and Oncology, University of Alabama at Birmingham Comprehensive Cancer Center, Birmingham, Alabama.

[8]Present address: Institute for Chemical Research, Kyoto University, Kyoto-Fu 611, Japan.

ISAO UNO (153), Institute of Applied Microbiology, University of Tokyo, Tokyo 113, Japan

LALITHA VAIDYANATHAN (165), Whitehead Institute for Biomedical Research, Massachusetts Institute of Technology, Cambridge, Massachusetts 02142

ROBERT WEINBERG (165), Whitehead Institute for Biomedical Research, Massachusetts Institute of Technology, Cambridge, Massachusetts 02142

MICHAEL WIGLER (153), Cold Spring Harbor Laboratory, Cold Spring Harbor, New York 11724

K. WIMAN (127), Molecular Biology and Virology Program of the Graduate School, Memorial Sloan-Kettering Cancer Center, New York, New York 10021

F. WONG-STAAL (183), Laboratory of Tumor Cell Biology, National Cancer Institute, Bethesda, Maryland 20205

Editor's Foreword

This volume reports the proceedings of the seventh Bristol-Myers Symposium on Cancer Research. The third and sixth symposia (from Stanford and Milan, respectively) were primarily oriented toward clinical treatment. The first, second, fourth, and fifth volumes described basic subjects like effects of drugs on the cell nucleus (Baylor), molecular targets for drug action (Yale), tumor cell heterogeneity (Hopkins), and chromosomes as they relate to cancer (Chicago).

The title of this symposium at the Memorial Sloan-Kettering Cancer Center is "Genetics, Cell Differentiation, and Cancer." Thus, it covers many topics introduced at the University of Chicago symposium held in 1982. It is extraordinary to reflect how much basic research has accumulated in the intervening two years, an indication of the accelerating pace of basic research in cancer.

Among many other themes, the current volume highlights the importance of protein factors in the growth, differentiation, and control of cells. It is singularly apt that tumor necrosis factor, first characterized at the Memorial Sloan-Kettering Cancer Center a dozen years ago, has now been cloned by a number of laboratories around the world and is entering clinical trial. The finding at Sloan-Kettering and elsewhere, that γ interferon is highly synergistic with TNF, suggests avenues for clinical exploration. Finally, it is not far-fetched to speculate that as many as a half dozen lymphokines might be used together in patient treatment, involving such diverse elements as interleukin II, tumor inhibitor factor, transforming growth factor, lymphotoxin, etc.

Maxwell Gordon
Series Editor

Foreword

A century ago, the public's perception of cancer was still medieval; it was considered a disease of shame. When we think of impressive advances in prevention and treatment during the past few decades, we must also be thankful for the changes in public perception and acceptance that now permit us to discuss cancer with relative freedom and candor.

The intense public interest in news about all aspects of cancer research extends to basic science as well as to clinical applications. In recent years, the public has learned of stunning advances in the understanding of how normal cells grow and develop and of what happens when the controls to that growth and development break down. Never before in the history of molecular biology have there been so many ideas to be tested. Never before have there been so many techniques to evaluate them.

But there is another perspective as well. In the weeks just before this symposium was held, the British scientific journal *Nature* reversed its earlier editorial position, saying it had been "overly optimistic" in predicting the speed with which oncogene research would lead to an understanding of the cause of some forms of cancer. And it is indeed the case that although remarkable progress has been made in some areas, the problem of solving the cancer riddle remains a vast one.

It was to these unsolved problems that the distinguished scientists participating in the seventh annual Bristol-Myers Symposium on Cancer Research addressed themselves. The symposium series is an integral part of the $8.34 million, no-strings-attached program of unrestricted grants for cancer research our company has sponsored since 1977 at 17 institutions in the United States and abroad.

The symposium organized by Memorial Sloan-Kettering Cancer Center had special meaning to us because it took place during Memorial Sloan-Kettering's centennial year and was held in New York, the city that has been our company's corporate headquarters for more than 80 years. With the publication of these proceedings, we are pleased that the important insights shared with the investigators attending the 1984 symposium can now be shared with a broader audience.

Richard L. Gelb
Chairman of the Board
Bristol-Myers Company

Preface

The Memorial Sloan-Kettering Cancer Center celebrated its Centennial Year in 1984. As the major scientific meeting in this hundredth year, the symposium "Genetics, Cell Differentiation, and Cancer," the seventh annual symposium on cancer research, was organized by the Center and sponsored by the Bristol-Myers Company. This volume is composed of papers delivered by scientists who participated in this conference.

Remarkable progress has been made in our understanding of the nature of carcinogenesis and, in particular, of the role of protooncogenes and oncogenes in the transformation of normal to abnormal cells. As these proceedings demonstrate, work in several areas, including studies on DNA replication, protein synthesis, embryologic development, DNA and RNA viruses, growth factors, and receptors, are coming together to provide us with an understanding of the process of transformation on a molecular and cellular level. This volume provides elegant testimony to the importance of studies on prokaryotes, simple eukaryotic organisms such as yeast, and *Drosophila* in providing insight into the nature of cancer. The first series of papers in these proceedings reports on studies of DNA replication, DNA structure, RNA synthesis, and control of gene expression. Subsequent papers review recent research on oncogenes and their role in inducing malignant growth.

The organizing committee for this symposium comprised Drs. Dorothea Bennett, Bayard D. Clarkson, William Hayward, Samuel Hellman, Paul A. Marks, Malcolm Moore, Lloyd J. Old, and Richard A. Rifkind. We are grateful to all of the participants who made this symposium and this volume which records its proceedings such a memorable scientific event.

We are particularly appreciative of the support of Mr. Richard Gelb, Chairman of the Board of the Bristol-Myers Company and his colleagues, in particular Mr. Harry Levine, Ms. Ann Wyant, and Ms. Kathryn Bloom.

Ms. Suzanne Rauffenbart and Ms. Suzanne Emery of the Memorial Sloan-Kettering Cancer Center provided excellent staff work in all details related to the organization of the symposium. We also thank Ms. Helene Friedman for her expert assistance in the preparation of these proceedings.

Paul A. Marks

Abbreviations

abl, Abelson murine leukemia virus (mouse)
ad, adenovirus
AIDS, acquired immunodeficiency disease syndrome
ALL, acute lymphoblastic leukemia
AML, acute myelogenous leukemia
ALV, avian leukosis virus
ARC, AIDS-related complex
ATTB, bacterial attachment site
ATL, adult T-cell leukemia
att, attachment site
ATP, adenosine triphosphate
attP, phage attachment site

BCGF, B-cell growth factor
*bgl*I, restriction enzyme
BLV, bovine leukemia virus

c, cellular
CAT, chloramphenicol acetyl transferase
CFUe, colony-forming unit for erythropoiesis
CMGF, cellular myeloid growth factor
CML, chronic myelogenous leukemia
CMP, cytidine monophosphate
c-onc, cellular or protooncogenes
CSA, colony-stimulating factor
CTP, cytidine triphosphate

d, deoxy
DBP, DNA binding protein

DIA, differentiation inducing activity
DMSO, dimethylsulfoxide
*eco*RI, restriction enzyme
EGF_2, epidermal growth factor
EGMA, eosinophil growth and maturation activity
ENV, viral envelope protein
erb, erythroblastosis virus
erbA, avian erythroblastosis virus (chicken)
erbB, avian erythroblastosis virus (chicken)
ets, E26 virus (chicken)

FAF, fibroblast activating factor
fes, ST feline sarcoma virus (cat)
fgr, Gardner–Rasheed feline sarcoma virus (cat)
fms, McDonough feline sarcoma virus (cat)
fos, FBJ osteosarcoma virus (mouse)
fps, Fujinami sarcoma virus (chicken)

gag, segment of human retroviral DNA
GTP, guanosine triphosphate

Ha, Harvey
Ha-*ras*, Harvey murine sarcoma virus (rat)
HLA, histocompatibility antigen
HmBA, hexamethyl bis acetamide
HTLV, human T-cell leukemia virus

IL-2, interleukin-2

kdal, kilodalton
Ki, Kirsten
Ki-*ras*, Kirsten murine sarcoma virus (rat)

λ, integrative recombination
LMEF, leukocyte migration enhancing factor
LMIF, leukocyte migration inhibitory factor
lor, large open reading frame
LTR, long terminal repeat

MAF, macrophage activating factor
MELC, murine erythroleukemia cells

mil(mht), MH2 virus (chicken)
MMEF, macrophage migration enhancing factor
mos, Maloney sarcoma virus (mouse)
myb, avian myeloblastosis virus (chicken)
myc, MC29 myelocytomatosis virus (chicken)

N, human
neu, oncogene associated with neuroblastoma
N-*myc*, myeloblastosis virus
nus, *E. coli* genes

onc, oncogene

PDGF, platelet-derived growth factor
pol, segment of human retroviral DNA
pTP, terminal protein precursor

raf, 3611 Murine sarcoma virus (mouse)
ras, murine sarcoma virus
rel, reticuloendotheliosis virus (turkey)
rho, protein factor
ros, URII avian sarcoma virus (chicken)

Saf, phase attachment site mutant
SDS, sodium dodecyl sulfate
sis, simian sarcoma virus (woolly monkey)
ski, avian SKV770 virus (chicken)
SphI, restriction enzyme
src, Rous sarcoma virus (chicken)

Tau, *E. coli* termination factor
TCGF, T-cell growth factor (IL-2)
Tp, tissue protein
T + P, thymine triphosphate
TPA, 12-*O*-tetradecanoylphorbol 13-acetate

v, virus
v-onc, viral oncogene

yes, Y73 sarcoma virus (chicken)

Introduction: Genetics, Cell Differentiation, and Cancer*

PAUL A. MARKS

DeWitt Wallace Research Laboratory
Sloan-Kettering Institute
Memorial Sloan-Kettering Cancer Center
New York, New York

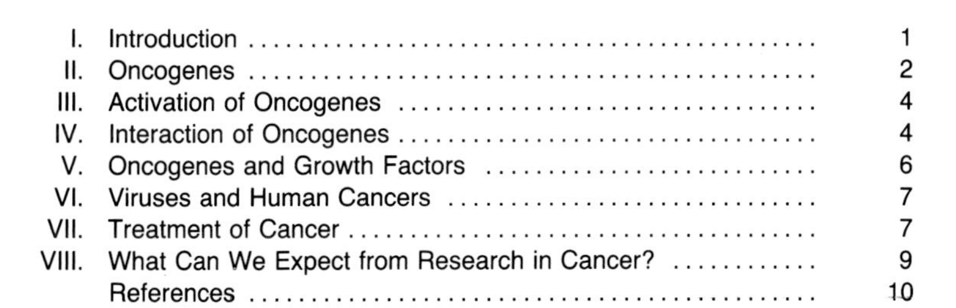

I. Introduction

During the past few years, there has been substantial progress in our understanding of the process of transformation of normal to malignant cells. These advances have generated widespread excitement over the opportunities that lie ahead in understanding the nature of the causes of many human cancers. Possibly the most important new discovery has been the cellular oncogenes. It seems likely that increased understanding of the biology of cancer will mean increased

*Research summarized in this review from the author's laboratory was supported, in part, by the National Cancer Institute (PO CA 31768 and CA 08748).

understanding of the biology of normal cell proliferation and differentiation. The organization of this conference on "Genetics, Cell Differentiation, and Cancer" reflects this concept, namely, that research on gene structure and gene expression in various prokaryotic and eukaryotic systems is essential for advancing our knowledge of the biology of cancer.

Until the 1970s, the direct study of the genes of organisms more complex than bacteria and viruses was markedly constrained by lack of experimental methods. The advent of recombinant DNA and other techniques has made it feasible to isolate genes responsible for particular functions and to determine their precise nucleotide structure (*1*). More to the point with respect to carcinogenesis, we are identifying genes that appear to play a role in causing human cancers and we are deciphering the controls that affect the expression of these genes (*2–4*).

II. Oncogenes

During the past few years, accumulated evidence has implicated more than 20 oncogenes—genes that appear to have a role in producing cancers in animals and causing the malignant transformation of cultured cells (Table I) (*2*). The evidence implicating oncogenes in the development of human cancers is still largely circumstantial. Most acutely transforming RNA tumor viruses have been shown to contain genomic sequences that appear to be responsible for the induction of neoplasia. These sequences have been termed viral oncogenes, or *v-onc*. The mechanism by which *v-onc* gene products mediate the transformation of normal to neoplastic cells still remains unclear. DNA sequences, so-called *c-onc* or protooncogenes, homologous to most of the known *v-onc* genes, have been identified in a variety of normal, uninfected cells, including human cells. The viral acquisition of protooncogenes is believed to have occurred by recombination events between the genome of the infecting retrovirus and that of the host cell. Protooncogenes have been detected in lower organisms such as yeast and drosophila, as well as in mice and humans. Genetic analysis of these lower organisms is already providing a means to analyze the properties of protooncogenes and define their role in cell growth and differentiation (*5*).

The structural similarity between cellular protooncogenes and viral oncogenes suggests that the former may possess potentially oncogenic material. At least four lines of evidence support the concept that cellular protooncogenes have oncogenic potential. First, *onc* genes *c-mos* and *c-Ha-ras,* when coupled to a

TABLE I

Oncogenes, Viruses, and Species of Origins

Oncogene	Virus and species of origin
abl	Abelson murine leukemia virus (mouse)
fes	ST feline sarcoma virus (cat)
fps	Fujinami sarcoma virus (chicken)
fgr	Gardner–Rasheed feline sarcoma virus (cat)
ros	UR II avian sarcoma virus (chicken)
src	Rous sarcoma virus (chicken)
yes	Y73 sarcoma virus (chicken)
erbB	Avian erythroblastosis virus (chicken)
fms	McDonough feline sarcoma virus (cat)
raf	3611 Murine sarcoma virus (mouse)
mil(mht)	MH2 virus (chicken)
mos	Maloney sarcoma virus (mouse)
rel	Reticuloendotheliosis virus (turkey)
sis	Simian sarcoma virus (woolly monkey)
Ha-ras	Harvey murine sarcoma virus (rat)
Ki-ras	Kirsten murine sarcoma virus (rat)
N-ras[a]	
myc	MC29 myelocytomatosis virus (chicken)
N-myc[a]	
erbA	Avian erythroblastosis virus (chicken)
ets	E26 virus (chicken)
ski	Avian SKV770 virus (chicken)
fos	FBJ osteosarcoma virus (mouse)
myb	Avian myeloblastosis virus (chicken)

[a] Origin: human neuroblastoma cell line, related to viral oncogene.

retroviral long terminal repeat sequence and transfected into NIH/3T3 cells, induce transformation (*6, 7*). Second, the avian leukosis virus, which lacks an identifiable oncogene, induces malignancy after a latent period (*8*). The mechanism appears to involve insertion of a viral genome adjacent to a cellular protooncogene, *c-myc,* which is thereby activated. Third, certain human tumor cell lines and fresh human tumors contain genes that induce a transformed phenotype when introduced into mouse NIH/3T3 cells, particularly *c-onc* genes known as *c-Ha-ras, c-Ki-ras,* and *c-N-ras* (*9–11*). Fourth, analysis of the expression of cellular oncogenes in fresh human tumors of patients with a number of different tumor types showed that more than one cellular oncogene was transcriptionally active in all of the tumors examined (Table II) (*12*). Taken together, these studies provide support for what has been thought by many cancer researchers for some time, namely, that gene changes contribute to the development of many cancers.

III. Activation of Oncogenes

The present evidence suggests that one of several mechanisms may lead to the activation of cellular protooncogenes. It has been known for several years that specific chromosomal abnormalities are associated with certain kinds of cancer. Research on chromosomal abnormalities has now begun to merge with research on oncogenes. The translocation of normal cellular counterparts of the oncogenes to new chromosomal locations seems to be one mechanism for activation of the *c-onc* genes (*13, 14*). For example, the aberrant chromosomes seen in cells of human cancer, such as Burkitt's lymphoma, have been shown to contain a misplaced gene, *c-myc,* which has moved from its normal location on chromosome 8 into the region on chromosome 14, coding for the heavy chain of the immunoglobulin molecule (*15*). The rearranged *c-myc* gene has been shown to be actively transcribed.

Another mechanism implicated in the activation of *c-onc* genes has been the amplification of specific oncogenes as, for example, *c-myc* gene in human small cell carcinoma of the lung (*16*) and HL60, a human promyelocytic leukemia cell line (*17*), and *c-myb* gene in acute myelogenous leukemia (*18*). *N-myc* amplification appears to be a frequent occurrence in neuroblastoma cells and has been demonstrated in retinoblastoma (*19*).

Oncogene activation may also be a consequence of a structural alteration in the gene such as a point mutation, as in the instance of *c-ras* associated with a bladder carcinoma cell line (*20*).

IV. Interaction of Oncogenes

Clinical experience with cancer suggests that human cancers generally develop slowly, probably involving several steps accumulated over a period of time (*21*). Thus, the incidence of most cancers increases with the average age of the population. There is a lag period between exposure to a known or suspected carcinogen, such as ionizing radiation or cigarette smoking, and clinically evident cancer.

TABLE II

Expression of *c-onc* Genes in Human Cancers[a]

Type of tumor	*c-onc* Gene								
	abl	*fes*	*fos*	*fms*	*myb*	*myc*	*Ha-ras*	*Ki-ras*	*src*
Renal cell	0	0	++	++	0	+++	++	++	0
Ovarian adenoma	0	0	+++	±	0	++	++	++	0
Uterine adenoma	0	0	0	0	0	0	0	0	0
Germ-cell tumor	0	0	++	0	0	++	++	++	0
Colorectal	0	0	+++	±	0	++	++	++	0
Small bowel	0	0	+++	0	0	++	++	++	0
Pancreatic	0	0	++	++	0	+	++	0	0
Lung	0	++	++	+	0	++	++	++	0
Breast	0	+	++	+++	+	+++	++	+	0
Pelvic sarcoma	0	0	++	0	0	+	+	0	0
Rhabdomyosarcoma	0	0	++	0	0	+++	++	+	0
Thymoma	0	0	0	0	0	+	+	0	0
Hodgkin's	0	0	++	0	0	+	++	+	0
Non-Hodgkin's	0	0	++	0	+++	++	++++	±	0
Acute myelocytic leukemia	0	+++	++	±	++++	++	+	++	0
Chronic myelocytic leukemia	++	++	0	0	+++	+	+	+	+
Acute lymphatic leukemia	0	+++	++	++	0	+++	+	+	0
Lymphoma	0	0	+	0	0	++	+	+	+++

[a] Adapted from Slamon *et al.* (*12*).

These observations have been interpreted as suggesting that cancer involves more than one defect; defects occur independently and sequentially; and these defects accumulated over time lead to the transformation of a normal cell to a tumor. This clinical experience seems at variance with the observation that transfection of NIH/3T3 cells with a single oncogene causes malignant transformation of these cells. NIH/3T3 cells are an immortal cell line in culture and probably represent a partially transformed phenotype from normal fibroblast counterparts. Compatible with the clinical experience is the more recent evidence that transformation of normal cells freshly placed in culture appear to require the activation of two or more oncogenes (*22*). The nature and the number of events required for transformation are unknown.

V. Oncogenes and Growth Factors

Research in several separate areas including RNA viruses, growth factors, and membrane structure has suggested the possible nature of the action of at least certain oncogenes (*23*). It has been shown that a partial sequence of a subunit of platelet-derived growth factor, PDGF, is nearly identical to the oncogene of simian sarcoma virus, designated *sis* (*24*). This finding suggests that an oncogene may contribute to the malignant transformation of cells by making the cells independent of exogenous growth regulatory factors. This theory received further support with the demonstration that the viral oncogene *erb-B* appears to be derived from a gene closely related to the receptor for epidermal growth factor, EGF (*25*). The product of *erb-B* represents a fragment of the transmembrane cellular receptor for EGF. The discovery of the relationship between EGF receptor and *erb-B,* and between PDGF and *sis,* suggests that certain oncogenes may cause cells to develop autonomy from normal external control of growth as a consequence either of growth factor–like proteins being produced within the transformed cell or of an alteration in a growth factor receptor that makes the tumor cell sense it as constituitively activated.

As many as half of the known viral oncogenes have either phosphokinase activity or structural features of a kinase. The receptors for certain growth-stimulating agents, including the hormone insulin as well as EGF and PDGF, are kinases (*2*). Taken together, these studies provide a model for further exploration of the mechanisms by which transformed cells can achieve relaxed control of cell proliferation, an important feature of neoplasia. Such studies are likely to increase our understanding of the mechanisms determining normal as well as transformed cell proliferation and differentiation.

VI. Viruses and Human Cancers

Activation of oncogenes may be one point of interaction between genetic and environmental determinants of cancer. One of the environmental factors that may be involved in activation of cellular oncogenes may be viruses (*4*). Viruses may also carry into the cell a transforming gene contributing to the causation of cancer. Viruses, however, appear to be a relatively rare cause of human cancers.

A possible role of viruses in the etiology of human cancers is, of course, not a new idea; scientists have long sought evidence for human tumor viruses, since the first report in 1908 implicating a virus in the etiology of a chicken leukosis. Particularly exciting have been the independent reports of investigators in the United States and Japan (*26*), which provided evidence that a human T-cell leukemia virus (HTLV) appears to be etiologically associated with an adult T-cell leukemia–lymphoma. The HTLV virus would represent the first human RNA virus to be discovered as a possible causative factor in human neoplasia. There are other candidates for cancer-associated viruses, including the Epstein–Barr virus in Burkitt's lymphoma and nasopharyngeal carcinoma, hepatitis B virus in hepatoma, and papilloma viruses and herpes viruses in cervical cancers, of which are DNA viruses.

VII. Treatment of Cancer

Significant advances have been achieved over the past three decades in the treatment of many forms of cancer (*27*). According to the National Cancer Institute report, the 5-year relative survival rate for cancer patients was 49% for patients diagnosed with cancer between 1976 and 1981. For certain specific cancers, 5-year relative survival rates are even better: thyroid, 92%; testes, 86%; endometrium, 85%; melanoma of the skin, 80%; female breast, 74%; bladder, 73%; Hodgkin's disease, 73%, and prostate, 70%. Survival rates continue to be poor for some cancers, such as lung, pancreas, stomach, esophagus, brain, and liver.

This improvement in the treatment of cancers has occurred without the fundamental understanding that we are beginning to gain about the nature of carcinogenesis. Indeed, the period between 1950 and 1970 saw the introduction of a number of what remain among the most useful chemotherapeutic agents, developed to a major extent on an empirical basis. But we face limitations in treatment with cytotoxic agents, including the toxicity of these agents and development of drug resistance. These limitations raise concern that further substantial improvement in the efficacy of chemotherapeutic agents of the cytotoxic variety may be difficult to achieve. I must emphasize that nothing we know today would suggest that we can or should abandon our continued efforts to improve our therapeutic capabilities through the traditional modes of treatment.

In the treatment of cancers, new alternatives to the use of cytotoxic drugs are being developed. These include agents that affect normal growth and differentiation of cells, the so-called growth factors and inhibitors of growth factors. In addition, a number of relatively simple chemical entities have been identified that can induce transformed cells to express characteristics of normal differentiated cells and lose their unlimited capacity for growth (*28, 29*).

Studies by my colleagues and myself on cytodifferentiating agents have focused primarily on the elucidation of cellular and molecular effects associated with inducer-mediated terminal cell division of transformed cells (*28*). We discovered that a variety of relatively simple chemical compounds can induce murine erythroleukemia cells to differentiate with the loss of proliferative capacity and the expression of differentiated characteristics, such as α and β globin genes. Among a number of agents that were examined, hexamethylene bisacetamide [HMBA, $H_3C{-}C(O){-}NH{-}(CH_2)_6{-}NH{-}C({=}O){-}CH_3{}^{r}$ was found to be particularly effective in inducing erythroleukemia cell differentiation. HMBA and related polar planar compounds can induce the expression of differentiated characteristics and the loss of proliferative capacity in a number of transformed cell lines. Among the cell lines induced by these agents are transformed cells of various species, including mouse, rat, dog, and human, derived from various cell types by chemical and viral transformation as well as lines that develop "spontaneously" (Table III) (*30*). There are types of tumor cells for which this is not the case, and we do not know what characteristics can be employed to predict which cells will be inducible by these agents. In general, it seems that inducer-sensitive transformed cell lines are blocked in the normal pathway of differentiation of these cells. Our studies have established that the inducers cause the modulation in expression of a number of specific genes that control differentiated functions, including loss of proliferative capacity. However, the mechanism by

TABLE III

Transformed Cells Induced to Differentiate by HMBA and/or DMSO[a,b]

Murine erythroleukemia
HL60, human promyelocytic leukemia
Rat mammary tumor cells
Canine kidney epithelial carcinoma
Mouse embryonal carcinoma
Mouse neuroblastoma
Human glioblastoma multiforme
Mouse teratocarcinoma
Rat LB myoblast
Human lymphoma cells
Mouse liver tumor cells
Human melanoma cells
Human colon cancer cells

[a] For detailed references, see Marks and Rifkind (*30*).

[b] HMBA, Hexamethylene bisacetamide; DMSO, dimethylsulfoxide.

which these chemical inducers reverse the malignant phenotype is not understood and remains a subject of further research. The use of such cytodifferentiating agents, as single or multiple agents, possibly with cytotoxic agents, is an area currently being evaluated both in animal models and in clinical trials. These studies are complex, but the results, while difficult to predict, may represent important contributions to new approaches to therapy.

VIII. What Can We Expect from Research in Cancer?

No one believes that an understanding of the mechanisms of carcinogenesis implies a cure, let alone the prevention of cancer. Indeed, for the near term, progress in the care of cancer patients is very likely to contribute to a far better outlook for many patients with cancer, while contributing to escalating costs of health care, as an increasing number of patients become affected with chronic diseases.

We should anticipate that new technologies will transform clinical practices, as well they should. One need only cite the use of biological response modifiers and cytodifferentiation chemicals; diagnostic magnetic resonance imaging and monoclonal antibodies; preventative dietary modification; and the psychosocial need to support cancer patients and their families as they face their illness. The implication of our new understanding of carcinogenesis is not that cancer will fade away in the next several years or even the next decades. Rather, what is likely to emerge are new approaches to earlier diagnoses of cancers, better techniques for treating them, and more effective strategies for preventing cancer—techniques that confront the fundamental causes of these diseases.

It would be misleading to believe that we are likely to develop a "magic bullet" that will cure all cancers or a vaccine to prevent all cancers. The evidence indicates that cancer is not a single disease, that the nature of cancer is such that it may be an integral part of living—the interaction between our genes and our environment. It is therefore unlikely that the solution to the cancer problem will be analogous to that achieved for infectious diseases, such as polio. It is also unlikely that we will eradicate cancer as we have eradicated smallpox. It *is* likely, however, that we will continue to increasingly control cancer and to become more effective in preventing some cancers and in curing many others.

References

1. T. Maniatis, E. F. Fritsch, and J. Sambrook (eds.), "Molecular Cloning." Cold Spring Harbor Laboratory, Cold Spring Harbor, New York, 1982.
2. J. M. Bishop, *Ann. Rev. Biochem.* **52,** 301 (1983).
3. R. A. Weinberg, *Sci. Am.* **249,** 126 (1983).
4. W. S. Hayward, B. G. Neel, and S. M. Astrin, *in* "Advances in Viral Oncology" (G. Klein. ed.), Vol. 1, p. 207. Raven, New York, 1982.
5. H. Hoffman-Falk, P. Einat, B.-Z. Shilo, and F. M. Hoffman, *Cell* **32,** 589 (1983).
6. D. G. Blair, M. Oskarsson, T. G. Wood, W. L. McClements, P. J. Fischinger, and G. G. Vande Woude, *Science* **212,** 941 (1981).
7. D. DeFeo, M. A. Gonda, H. A. Young, E. H. Chang, D. R. Lowy, E. M. Scolnick, and R. W. Ellis, *Proc. Natl. Acad. Sci. U.S.A.* **78,** 3328 (1981).
8. W. S. Hayward, B. G. Neel, and S. M. Astrin, *Nature (London)* **290,** 475 (1981).
9. C. Shih, L. C. Padhy, M. Murray, and R. A. Weinberg, *Nature (London)* **290,** 261 (1981).
10. C. J. Der, T. G. Krontiris, and G. M. Cooper, *Proc. Natl. Acad. Sci. U.S.A.* **79,** 3637 (1982).
11. T. G. Krontiris and G. M. Cooper, *Proc. Natl. Acad. Sci. U.S.A.* **78,** 1181 (1981).
12. D. J. Slamon, J. B. deKernion, I. M. Verma, and M. J. Cline, *Science* **224,** 256 (1984).
13. J. D. Rowley, *Nature (London)* **301,** 290 (1983).
14. J. J. Yunis, *Science* **221,** 227 (1983).

15. J. Erikson, A. Ar-Rushdi, H. L. Drwinga, P. C. Nowell, and C. M. Croce, *Proc. Natl. Acad. Sci. U.S.A.* **80,** 820 (1983).
16. C. D. Little, M. M. Nau, D. N. Carney, A. F. Gazdar, and J. D. Minna, *Nature (London)* **306,** 194 (1983).
17. R. Dalla-Favera, F. Wong-Staal, and R. C. Gallo, *Nature (London)* **299,** 61 (1982).
18. P. G. Pelicci, L. Lanfrancone, M. D. Brathwaite, S. R. Wolman, and R. Dalla-Favera, *Science* **224,** 1117 (1984).
19. G. M. Brodeur, R. C. Seeger, M. Schwab, H. E. Varmus, and J. M. Bishop, *Science* **224,** 1121 (1984).
20. E. P. Reddy, R. K. Reynolds, E. Santos, and M. Barbacid, *Nature (London)* **300,** 149 (1982).
21. P. A. Marks, *Blood* **58,** 415 (1981).
22. H. Land, L. F. Parada, and R. A. Weinberg, *Nature (London)* **304,** 596 (1983).
23. C-H Heldin and B. Westermark, *Cell* **37,** 9 (1984).
24. M. D. Waterfield, G. T. Scrace, N. Whittle, P. Stroobant, A. Johnsson, A. Wasteson, B. Westermark, C-H Heldin, J. S. Huang, and T. F. Deuel, *Nature (London)* **304,** 35 (1983).
25. J. Downward, Y. Yarden, E. Mayes, G. Scrace, N. Totty, P. Stockwell, A. Ullrich, J. Schlessinger, and M. D. Westerfield, *Nature (London)* **307,** 521 (1984).
26. R. C. Gallo and F. Wong-Staal, *Cancer Res.* **44,** 2743 (1984).
27. V. T. DeVita, Jr., "Cancer Patient Survival Statistics." Report of the National Cancer Institute/Office of Cancer Communications. Washington, D. C., November 26, 1984.
28. P. A. Marks, T. Murate, T. Kaneda, J. Ravetch, and R. A. Rifkind, *in* "Mediators in Cell Growth and Differentiation" (R. J. Ford and A. L. Maisel, eds.), p. 327. Raven, New York, 1985.
29. M. B. Sporn and A. B. Roberts, *Cancer Res.* **43,** 3034 (1983).
30. P. A. Marks and R. A. Rifkind, *Cancer* **54,** 2766 (1984).

PART I

Gene Structure and Expression

1

Synthesis of Adenoviral DNA with Purified Proteins

JERARD HURWITZ, SAMIT ADHYA, JEFFREY FIELD, RICHARD GRONOSTAJSKI, RONALD A. GUGGENHEIMER, MARK KENNY, JEFF LINDENBAUM, AND KYOUSUKE NAGATA

Department of Molecular Biology and Virology
Memorial Sloan-Kettering Institute for Cancer Research
New York, New York

I. Introduction

Detailed analyses of DNA replication have depended upon the use of model systems capable of being manipulated *in vitro* (*1*). To date, of the available systems in eukaryotes, the adenovirus replication system is the only one in which the entire replication process can be carried out *in vitro* (*2, 3*). The results obtained in a number of laboratories have contributed to our understanding of this pathway. The proteins essential for Ad DNA replication have been isolated and models explaining both *in vivo* and in *in vitro* DNA synthesis have been proposed (*4–6*).

The adenovirus genome (Ad DNA-pro) is a linear duplex DNA 35,000 bp (base pairs) long with a 5- kilodalton (kdal) protein (TP) covalently linked to each 5′ end. Five different proteins essential for the synthesis of full length Ad DNA pro have been isolated (*7*). Three of these, the 72-kdal Ad DNA Binding protein (Ad DBP), the 140-kdal Ad DNA polymerase (Ad Pol), and the 80-kdal

GENETICS, CELL DIFFERENTIATION, AND CANCER

TABLE I

Requirement for Ad DNA-pro Replication with Purified Protein

Addition	DNA Synthesis (pmol)
Complete[a]	24.5
Omit pTP–Ad Pol fraction, or Ad DBP	<0.10
Omit Nuclear factors I and II	1.12
Omit Nuclear factor I	1.28
Omit Nuclear factor II	7.40
Omit ATP	7.24
Omit Ad DNA pro	<0.10
Add Ad DNA (proteinase K treated)	0.20

[a] Complete: Ad DNA-pro (160 pmol of nucleotides), pTP–Ad Pol complex, Ad DBP, nuclear factor I, nuclear factor II, 4 dNTPs, and ATP.

precursor to the terminal protein (pTP) are viral gene products and are purified from infected HeLa cell extracts. The Ad Pol and the pTP are found as a complex (pTP–Ad Pol) but can be separated by glycerol gradient sedimentation in the presence of urea. The other two proteins, nuclear factor I and nuclear factor II, are host encoded and are purified from uninfected HeLa cell nuclear extracts. Table I shows the proteins and template requirements for *in vitro* replication of Ad DNA pro. Synthesis of DNA is markedly reduced upon omission of the viral coded Ad Pol, the Ad DBP, the pTP, and nuclear factor I. All synthesis is dependent on the presence of the template, Ad DNA pro, and the removal of the terminal protein by treatment with proteinase K reduces deoxynucleotide incorporation below detectable levels. The host protein, nuclear factor II, stimulates the reaction approximately fourfold, as does the addition of high levels of ATP (3 m*M*).

The size of the product formed in the complete system sedimented as full-length Ad DNA (34 S) in alkaline sucrose gradients. In the absence of nuclear factor II, the product was only 25% the length of intact Ad DNA. Thus, nuclear factor II stimulates DNA synthesis and is essential for the synthesis of full-length DNA.

The product synthesized in the *in vitro* reaction is covalently linked to protein. This can be readily demonstrated by adsorption and elution of the product from benzoylated–naphthoylated DEAE–cellulose (Table II). Protein-linked DNA is

TABLE II

BND–Cellulose Chromatography of Product[a]

Treatment of product	Protein-linked DNA	Unlinked DNA
None (1.2×10^6 cpm)	>95%	1.4%
Proteinase K (1.2×10^6 cpm)	<1.4%	>95%

[a] DNA products were synthesized as described to the legend of Table I and then subjected to chromatography on benzoylated–naphthoylated DEAE–cellulose and eluted as previously described (*18*).

eluted only after the addition of sodium dodecyl sulfate (SDS) and urea. When the product formed in the complete system was treated with proteinase K and then adsorbed to the column, it eluted as free DNA and no protein-linked material was detected. Thus, like DNA isolated from Ad virions, all DNA synthesized in the system is covalently linked to protein.

Two possible mechanisms could account for the protein-linked product. Protein could be added after initiation, or, as originally proposed by Rekosh *et al.* (*5*), the preterminal protein could prime DNA synthesis. In the latter mechanism one predicts a pTP–dCMP intermediate because the 5′ terminal nucleotide on each strand is dCMP. pTP–dCMP formation can be measured by SDS polyacrylamide gel electrophoresis of reaction mixtures in which the only deoxynucleoside triphosphate present is [α^{32}P]dCTP. In addition, limited elongation of the pTP–dCMP complex can be carried out in the presence of dATP, dTTP, dCTP, and ddGTP. The reaction proceeds only to the 26th nucleotide, where the first dGMP residue is incorporated into the growing chain. Autoradiographs of reaction mixtures in which only the initiation reaction occurs show a ^{32}P-labeled 80-kdal band, and autoradiographs of the limited elongation reaction show an 88-kdal band (a covalent complex between pTP and a 26 mer) in addition to the 80-kdal band. Table III shows the requirements for pTP–dCMP formation. Maximal complex formation required four of the five proteins essential for the synthesis of full-length Ad DNA pro; nuclear factor II did not affect the reaction. This is in agreement with the results presented above implicating nuclear factor II as an elongation factor.

Each of the proteins described above has been resolved and highly purified, and their properties are summarized in Table IV. The only DNA polymerase activity present in the system is the viral-coded enzyme. This enzyme is isolated as a stoichiometric complex with the 80-kdal pTP and can be resolved from this protein by glycerol gradient centrifugation in the presence of urea. The 140-kdal

TABLE III

Requirements for pTP–dCMP Synthesis with Purified Proteins

	pTP–dCMP formation (% of maximum)
Complete[a]	100
Omit pTP or Ad Pol or Ad DNA pro or $MgCl_2$	<5
Complete with Ad DNA (proteinase K)	<5
Omit nuclear factor I	14
Omit Ad DBP	30
Omit ATP	51
Complete plus nuclear factor II	100

[a] The complete reaction mixture contained Ad Pol, pTP, nuclear factor I, Ad DBP, ATP, Ad DNA pro, $MgCl_2$, and dCTP; 100% incorporation represented 0.93 fmol of dCMP covalently linked to the pTP.

protein possesses DNA polymerase activity, while the 80-kdal protein is required for formation of the pTP–dCMP complex. No other DNA polymerase of eukaryotic or prokaryotic origin substitutes for the viral polymerase in supporting the initiation reaction. The role played by the pTP other than its participation in the initiation reaction is unknown. It is also clear that the Ad DNA terminal proteins are important in the reaction. We suspect that the terminal proteins play important roles governing the initiation reaction as well as in the maintenance of the circular structure of the Ad DNA pro.

The Ad DBP, in addition to binding to DNA, specifically interacts with the Ad DNA polymerase. An example of this is shown in Table V. In the presence of poly(dT)-oligo(dA) [or oligo(rA)] the synthesis of poly(dA) was totally dependent on Ad pol and Ad DBP. The *E. coli* DNA binding protein did not substitute for Ad DBP, nor did DNA polymerase α substitute for the Ad DNA polymerase. In addition, a high concentration of ATP (4 m*M*) stimulates the synthesis of poly(dA); similar results are obtained when the nonhydrolyzable ATP analogs AppNp and AppCp are substituted for ATP. These results, coupled with the absence of detectable hydrolysis of ATP to ADP or AMP during DNA synthesis, suggest that ATP acts as an effector in the reaction without being hydrolyzed. Another important feature of the interaction between Ad DBP and Ad DNA pol is the size of poly(dA) products formed; they range between 30 and 40 kb in length under conditions in which <1% of the primers are elongated, suggesting

TABLE IV

Properties of Enzymes Involved in Ad DNA Replication

- Ad Pol
  - 140-Kdal protein
  - associated with 80-Kdal pTP
  - Located between 18 and 22 map units on Ad genome
  - Distinct from DNA polymerase α, β, and γ
    - Only polymerase capable of linking dCMP to 80-Kdal pTP
    - Uses poly(dT) and oligo(rA) or oligo(dA) as primer templates
    - Polymerase activity with poly (dT)-oligo (dA) is dependent on Ad DBP
    - Polymerase is resistant to aphidicolin
    - Single-stranded specific 3′ → 5′-exonuclease copurifies with polymerase
- Precursor to terminal protein (pTP)
  - 80-Kdal protein
  - Becomes covalently linked to dCMP to initiate synthesis
  - Located at 28.9–23.5 map units on Ad genome
  - Associated with 140-Kdal Ad Pol but separates in urea
  - Processed to 55-Kdal TP after initiation reaction
- Ad DBP
  - 59-Kdal protein (72 Kdal on SDS–polyacrylamide gels)
  - Binds single-stranded DNA
  - Interacts specifically with Ad Pol to allow highly processive DNA synthesis
  - Located between 62 and 68 map units on Ad genome
  - Involved in both initiation and elongation
- Nuclear factor II
  - Host protein isolated from nuclear extracts
  - No effect on initiation reaction
  - Required for full-length Ad DNA synthesis: in its absence, chains grow to 10 kb in length and then stop
  - Copurifies with DNA topoisomerase I activity; ATP has no effect: changes linking number in steps of one
- Nuclear factor I
  - Host protein isolated from nuclear extracts
  - 47-Kdal protein
  - Stimulates formation of pTP–dCMP complex
  - Binds specifically to nucleotides 17–48 of Ad 5 DNA
  - Binds specifically to DNA sequences in HeLa genome

that the Ad Pol is processive enough to replicate the entire adenovirus genome in a single binding event. This system also displaces the nonreplicated strand during synthesis.

Genetic studies carried out by McDonough and Rekosh (*8*) suggest a specific interaction between the Ad DBP and the Ad Pol. They showed that Ad mutants

TABLE V

Poly(dT)–Oligo(dA)–Primed DNA Synthesis Using Ad DNA Pol[a]

Additions	Activity (%)
pTP + Ad Pol + Ad DBP	100
Omit Ad Pol	<2
Omit pTP	280
Omit Ad DBP	<2
Omit Ad DBP, omit Ad Pol	<2
Omit Ad Pol, omit pTP and DNA Pol α	<2
Omit Ad DBP, add *E. coli* SSB	<2

[a] Reaction mixtures (50 μl) contained 50 m*M* Tris-HCl, pH 7.9, 10 m*M* $MgCl_2$, 4 m*M* dithiothreitol, 10 μg bovine serum albumin, 0.4 μg poly(dT), 0.4 μg oligo(dA), 8 μ*M* [^{3}H]dATP (2500 cpm/pmol), 0.8 μg Ad DBP or *E. coli* SSB, and separated pTP and Ad Pol. In the above experiments, 100% represented 10.2 pmol of dAMP incorporated.

in which the Ad DBP was thermolabile supported DNA replication upon complementation with some human serotypes but not with all. Their results suggest a high degree of specificity in the interaction between Ad DBP and other proteins involved in replication. Our biochemical results support the genetic observations and indicate that the serotype-specific complementation may be at the level of the Ad Pol–Ad DBP interaction.

II. Role Played by Host Factors

Nuclear factor II is required for the synthesis of full-length Ad DNA. The purified protein possesses DNA topoisomerase I activity, an activity that copurified with the fraction supporting the synthesis of full-length Ad DNA. Nuclear factor II was replaced by eukaryotic topoisomerase I (isolated from HeLa cells or calf thymus). In all three cases (nuclear factor II and the two eukaryotic topoisomerase I activities), full-length Ad DNA synthesis was obtained with identical units of topoisomerase I activity. It is interesting that the type I DNA topoisomerase isolated from *E. coli* did not substitute for the eukaryotic enzyme.

The eukaryotic topoisomerase I has been shown to be a protein of 100 kdal (*9*); in contrast, nuclear factor II possesses an apparent molecular mass of 30 kdal. At present this discrepancy has not been resolved. While it is possible that eukaryotic cells contain different topoisomerases, it is equally possible that we have isolated a fragment that has suffered proteolytic attack but has retained its enzymatic activity.

The role played by topoisomerase I in the synthesis of full-length Ad DNA is unclear at present. The initiation and elongation of chains occur in the absence of topoisomerase I; when synthesis has ceased, the addition of topoisomerase I to such a reaction mixture results in an immediate response and synthesis of full-length Ad DNA. These observations indicate that the replication fork is arrested and blocked in its translocation through the chain. We suspect that the displaced strand interferes with the movement of the replication fork and topoisomerase I relieves such obstructions. Alternatively, interaction between the two terminal proteins may create a circular, topologically constrained template. When DNA preparations <10 kb long are used as templates, the synthesis of full-length products occurs in the absence of topoisomerase I.

The host protein, nuclear factor I, markedly increases initiation of Ad DNA synthesis. The properties of this host protein are also summarized in Table IV. Nuclear factor I has the remarkable ability to bind specifically to a sequence present within the first 50 nucleotides at either end of Ad DNA. The protein binds poorly to single-stranded DNA and binds selectively to duplex DNA. DNase ''footprinting'' experiments have revealed a binding site located between nucleotides 17–49 from the end. The stoichiometry of binding indicates that there are 2 mol of nuclear factor I bound per mole of binding site. Studies carried out with plasmids containing the left end of Ad DNA (pLA1) clearly indicate that two regions, lying within the first 50 nucleotides, are essential for DNA replication. One is a highly conserved region between nucleotides 9–18 found in all human serotypes (*6, 10*) while the other is the factor I binding site (Fig. 1). Alteration of these sites reduced DNA replication below detectable levels.

The role of nuclear factor I in DNA synthesis is now under investigation. We have noted that DNase footprinting in the presence of nuclear factor I results in increased nuclease cutting at sites adjacent to the factor I binding site; similar results have been obtained with endonucleases specific for single-stranded DNA. These findings suggest that nuclear factor I binds to DNA, producing single-stranded region in the areas before and after the binding sites.

We have detected nuclear factor I DNA binding sites in HeLa DNA. The DNA

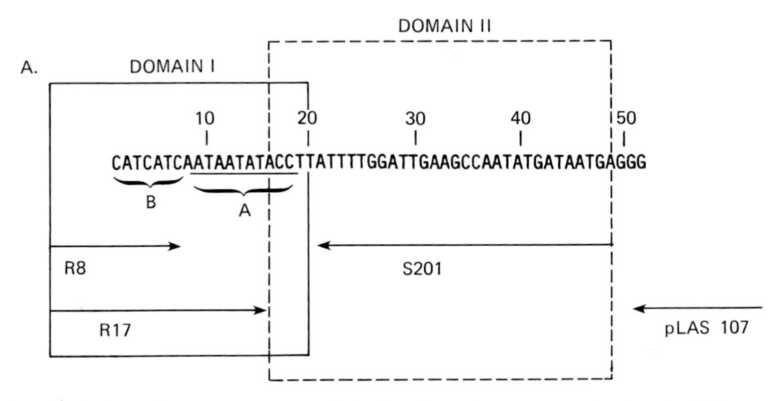

Fig. 1. Schematic representation of the two DNA domains required for Ad DNA replication. Domain I can be divided into two areas: A, a 10-bp core sequence (underlined) that the pTP may recognize and bind and B, a spacer region separating the core sequence and the terminal dG/dC base pair. Arrows represent the boundaries of the deletion plasmids used to define both domains; for further details, see Ref. *18*.

sequences protected from DNase attack in the HeLa DNA sites have been determined (Fig. 2) and compared to the similar site in pLA1 DNA. A common sequence, as indicated, was noted in all these DNAs. Recently, Borgmeyer *et al.* (*11*) and Siebenlist *et al.* (*12*) have detected binding sites in DNA sequences thought to be involved in the regulation of transciption in the lysozyme and the *c-myc* genes. These results suggest that nuclear factor I binding sites may play an important role in regulating RNA synthesis and possibly DNA synthesis.

A model of Ad DNA replication that explains many of the results is presented in Fig. 3. We suggest that the structure of Ad DNA-pro is circular. Electron micrographs of Ad DNA-pro show substantial numbers of circular structures.

aatataccttATTTTGGATTGAAGCCAATATGATAATGaggggtggagt--------Ad 5

ctaaatctAGAACCTGGGCACAGTGCCAAGTTGTAGccaattcccaggc-------FIB-1

gggcctGCTAGGTCTGGCTTTGGGCCAAGAGCCgctgcgtttcag-----------FIB-2

agatccagATCCTCTGGAATTGTGCCAAATGTcctcatgccacttgcaca------FIB-3

Consensus sequence = CTGG(N_{6-7})GCCAA

Fig. 2. Nucleotide sequence of DNA regions protected from DNase I digestion by nuclear factor I.

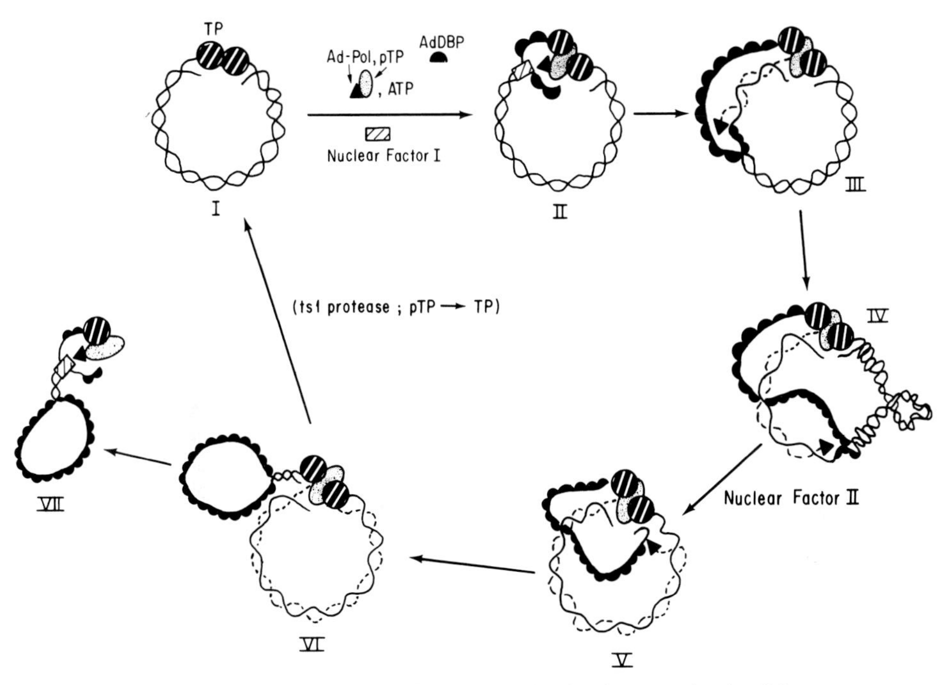

Fig. 3. Proposed model of Ad DNA-pro replication (see text for details).

Removal of the terminal proteins (with proteinase K) results in linear duplex structures and no circular forms (*13, 14*). In the model presented, four of the five proteins required for replication are involved in the initiation of replication and lead to the formation of products as long as 10 kb. The parental strand containing the terminal protein is displaced, leading, eventually, to the product IV in Fig. 3. The reaction ceases but can be resumed upon the addition of topoisomerase I, leading to the displacement of an intact single strand (V), which can form a panhandle structure as shown in VII. The duplex structure (VI) contains one newly synthesized strand covalently linked to an 80-kdal protein. The pTP is processed to a 55-kdal protein by a viral-coded protease (*15*).

A panhandle structure (VII) can form due to the presence of complementary regions at each end (*16*). This results in a duplex region of 100–200 nucleotides (depending on the serotype) that contains all of the signals essential for the initiation of Ad DNA replication. The difference between the replication of structure VII and Ad DNA-pro (structure I) occurs after the replication fork traverses the panhandle region. The fork then must replicate a single-stranded template of more than 34,000 nucleotides. This type of replication can be readily carried out by the Ad polymerase in the presence of Ad DBP.

The above model is basically a modified rolling circle, in which single-strand displacement occurs concomitant with synthesis. What happens to the proteins upon termination of one round of DNA synthesis is unknown and remains to be elucidated.

Ad replication has proved useful as a system for elucidating the way in which numerous enzymes interact to synthesize this genome. Of particular interest is identifying the role in host cell nucleic acid metabolism played by nuclear factor I and nuclear factor II. In addition, the displacement synthesis catalyzed by the Ad DBP–Ad Pol complex may prove to be a general mechanism for eukaryotic DNA unwinding as no eukaryotic helicases have been reported. Recently, Ariga and Sugano (*17*) have reported an *in vitro* replication system dependent on SV40 DNA. This system, which initiates with RNA primers and is bidirectional, probably utilizes more host-encoded proteins than does the Ad replication system described here. Biochemical analysis of the SV40 DNA system should provide further insight into the replication of host DNA.

References

1. A. Kornberg, "DNA Replication." Freeman, San Francisco, California, 1980.
2. M. D. Challberg and T. J. Kelly, Jr., *Proc. Natl. Acad. Sci. U.S.A.* **76,** 655 (1979).
3. K. Nagata, R. A. Guggenheimer, and J. Hurwitz, *Proc. Natl. Acad. Sci. U.S.A.* **80,** 6177 (1983).
4. R. L. Lechner and T. J. Kelly, Jr., *Cell* **12,** 1007 (1977).
5. D. M. K. Rekosh, W. C. Russell, A. J. D. Bellett, and A. J. Robinson, *Cell* **11,** 283 (1977).
6. B. W. Stillman, *Cell* **35,** (1983).
7. K. Nagata, R. A. Guggenheimer, and J. Hurwitz, *Proc. Natl. Acad. Sci. U.S.A.* **80,** 4266 (1983).
8. J. S. McDonough and D. M. K. Rekosh, *Virology* **120,** 383 (1982).
9. L. F. Liu and K. G. Miller, *Proc. Natl. Acad. Sci. U.S.A.* **78,** 3487 (1981).
10. M. Shinagawa and R. Padmanabhan, *Proc. Natl. Acad. Sci. U.S.A.* **77,** 3831 (1980).
11. U. Borgmeyer, J. Nowock, and A. E. Sippel, *Nucleic Acid Res.* **12,** 4295 (1984).
12. U. Siebenlist, L. Hennighausen, J. Battey, and P. Leder, *Cell* **37,** 381 (1984).
13. A. J. Robinson, H. B. Younghusband, and A. J. D. Bellett, *Virology* **56,** 54 (1973).
14. A. J. Robinson and A. J. D. Bellett, *J. Virol.* **15,** 458 (1975).
15. J. Weber, *J. Virol.* **17,** 462 (1976).
16. J. Wolfson and D. Dressler, *Proc. Natl. Acad. Sci. U.S.A.* **69,** 3054 (1972).
17. H. Ariga and S. Sugano, *J. Virol.* **48,** 481 (1983).
18. R. A. Guggenheimer, B. W. Stillman, K. Nagata, F. Tamanoi, and J. Hurwitz, *Proc. Natl. Acad. Sci. U.S.A.* **81,** 3069–3073 (1984).

Left-Handed Z-DNA and Chromatin

ALEXANDER RICH

Department of Biology
Massachusetts Institute of Technology
Cambridge, Massachusetts

I. Introduction

Since 1953 we have known that can DNA exist in two different right-handed double-helical forms, A- and B-DNA (*1*, *2*). Recently we have learned that it can also adopt a left-handed double-helical conformation (*3*) (reviewed in Ref. *4*). The right-handed conformations can accommodate any sequence of nucleotides and the external form of the molecule does not change significantly with different sequences. However, the left-handed form is more stable in sequences that have

GENETICS, CELL DIFFERENTIATION, AND CANCER

purine–pyrimidine alternation. When it deviates from this, the external form of the molecule changes (*5*).

Only recently has it been possible to study the molecular structure of DNA by single cyrstal X-ray analysis. This was made possible by advances in synthetic organic nucleotide chemistry. Oligonucleotides can now be made with a defined sequence in quantities sufficient for crystallization. Single crystals often diffract X rays at or near atomic resolution (~1 Å). Such crystal structures provide a wealth of detail in contrast to the significantly more limited results from DNA fiber X-ray diffraction analyses. A double-helical fragment was first visualized at atomic resolution in 1973 with crystalline dinucleoside monophosphates, which showed right-handed base-paired RNA fragments (*6, 7*). The left-handed DNA structure was discovered in 1979 in a crystalline hexanucleoside pentaphosphate with the sequence d(CpGpCpGpCpG) (*3*). That crystal diffracted to 0.9-Å resolution and hence all atoms were visualized, including solvents and salts.

II. What Happens to DNA in Different Conformations?

We would like to know what induces conformational changes in DNA. Some conformations are more stable in a particular environment. How is this controlled and can we show that different conformations are found in chromatin *in vivo* as well as *in vitro*? Here we address some of these issues for left-handed DNA.

The double helical hexanucleoside pentaphosphate molecules were found aligned along the axis of the crystal in the left-handed DNA crystal structure (*3*), and thus the molecule appeared to be a continuous helix. A van der Waals diagram of the structure is shown in Fig. 1 as it is found in the crystal lattice. The backbone has a zigzag arrangement and the molecule was called Z-DNA. Three molecules are drawn and there is a continuity of base pair stacking along the entire helical axis. In contrast to B-DNA, Z-DNA has one deep helical groove, which is formally analogous to the minor groove of B-DNA. The asymmetric unit in Z-DNA is a dinucleotide compared to a mononucleotide in B-DNA. The heavy line in Fig. 1 drawn between adjacent phosphate groups shows their zigzag organization in Z-DNA. This is due to the existence of two different nucleotide conformations in Z-DNA. The conformation of deoxyguanosine in Z-DNA and B-DNA is shown in Fig. 2. In B-DNA, all of the bases have the *anti* conformation and a C-2′ *endo* pucker of the deoxyribose ring. In the Z-DNA crystal with

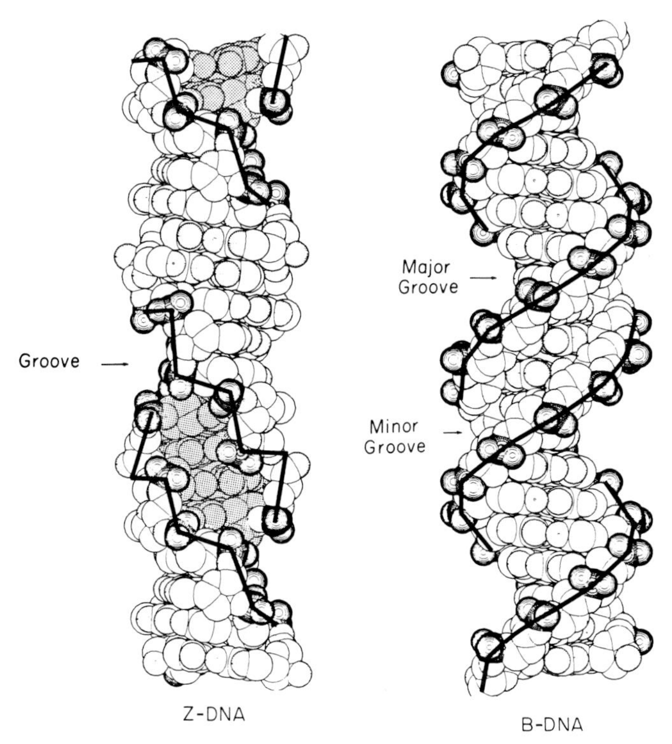

Fig. 1. Van der Waals diagrams of Z-DNA and B-DNA. The irregularity of the Z-DNA backbone is illustrated by the heavy lines which go from phosphate to phosphate residue along the chain. Z-DNA is shown as it appears in the hexamer crystal (*3*). In contrast, B-DNA has a smooth line connecting the phosphate groups and two grooves, neither of which extends to the axis of the helical molecule.

alternating cytosine and guanine residues, the deoxycytidines all have the *anti* conformation while the deoxyguanosines are all *syn*. Every other residue in a Z-DNA molecule has the *syn* conformation. Thus there is a dinucleotide repeat in Z-DNA due to the alternations of *anti* and *syn* conformations.

There are 12 base pairs per turn of the helix in Z-DNA, a pitch of 44.6 Å, and a diameter near 18 Å. In contrast, B-DNA typically has 10.5 base pairs per helical turn, a helical pitch of 34 Å, and a diameter of 20 Å. Z-DNA is thus slightly slimmer than B-DNA and has more base pairs per helical turn. Both Z-DNA and B-DNA are built out of two antiparallel polynucleotide chains held together by Watson–Crick base pairs. However, the base pairs in the two struc-

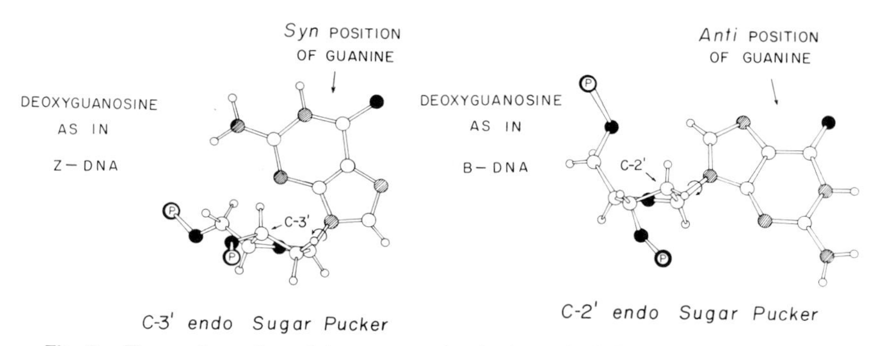

Fig. 2. The conformation of deoxyguanosine is shown in B-DNA and in Z-DNA. In the sugar, the plane defined by C-1′–O-1′–C-4′ is horizontal. Atoms lying above this plane are in the *endo* conformation. The C-3′ is *endo* in Z-DNA while in B-DNA the C-2′ is *endo.* Z-DNA has guanine in the *syn* position, in contrast to the *anti* position in B-DNA. A curved arrow around the glycosyl carbon–nitrogen linkage indicates the site of rotation.

tures have a different relationship to the sugar phosphate backbone. In Fig. 3, DNA is drawn unwound in the familiar ladder representation with the planar bases shown as flat plates. The base pairs must flip over in converting a section of B-DNA to Z-DNA so that they are upside down relative to their initial orientation. This flipping is brought about by a rotation of every other residue about its glycosyl bond, from the *anti* to the *syn* conformation. In the cyrstal structure of Fig. 1, each guanine goes to the *syn* conformation. For the pyrimidines, both the base and the sugar rotate. It is the rotation of every other sugar along the chain that produces the zigzag backbone conformation.

In Z-DNA, the base pairs are located away from the center of the molecule such that the guanine imidazole ring is found at the periphery. In B-DNA, the base pairs are located on the helical axis. These are shown in the projections down the helix axis of both Z-DNA and B-DNA in Fig. 4. The size and depth of the helical groove in Z-DNA is shown in Fig. 5. The groove extends almost to the helical axis.

III. Other Sequences Form Z-DNA Crystals

A DNA fragment with the sequence d(*CpGpTpAp*CpG) crystallized as Z-DNA and was resolved to 1.2-Å resolution (*8*). The cytosine residues have either methyl or bromine atoms attached to the C-5 position, which is known to sta-

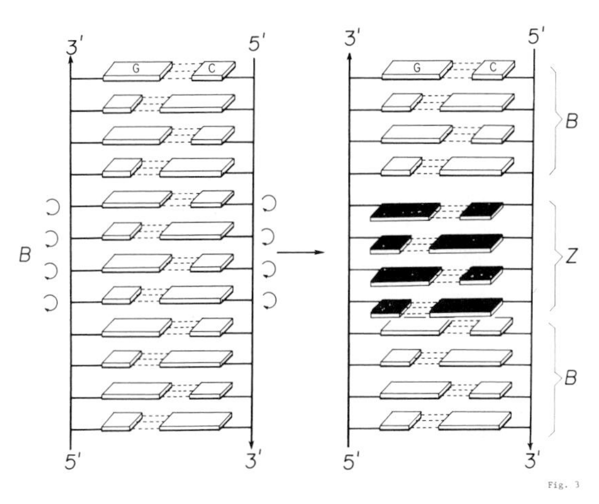

Fig. 3. The change in topological relationship is shown when a 4-bp segment of B-DNA converts to Z-DNA. The conversion is accomplished by rotation or flipping over the base pairs as indicated by the curved arrows. Rotation of alternate residues about the glycosylic bond puts them into the *syn* conformation.

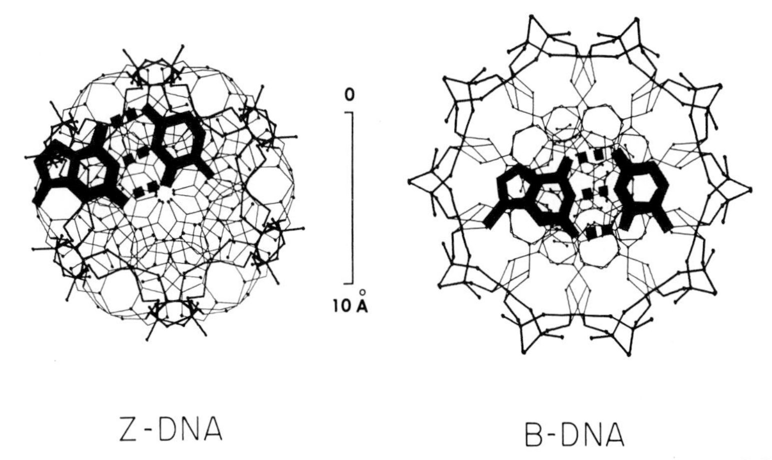

Fig. 4. Skeletal diagrams of projections down the helix axis of both Z-DNA and B-DNA. The sugar–phosphate backbone is drawn with somewhat heavier lines. One guanine–cytosine base pair is accentuated in the helix to show the relative difference in their position in B-DNA and Z-DNA. In Z-DNA the base pairs are located near the periphery of the molecule in contrast to their central position in B-DNA.

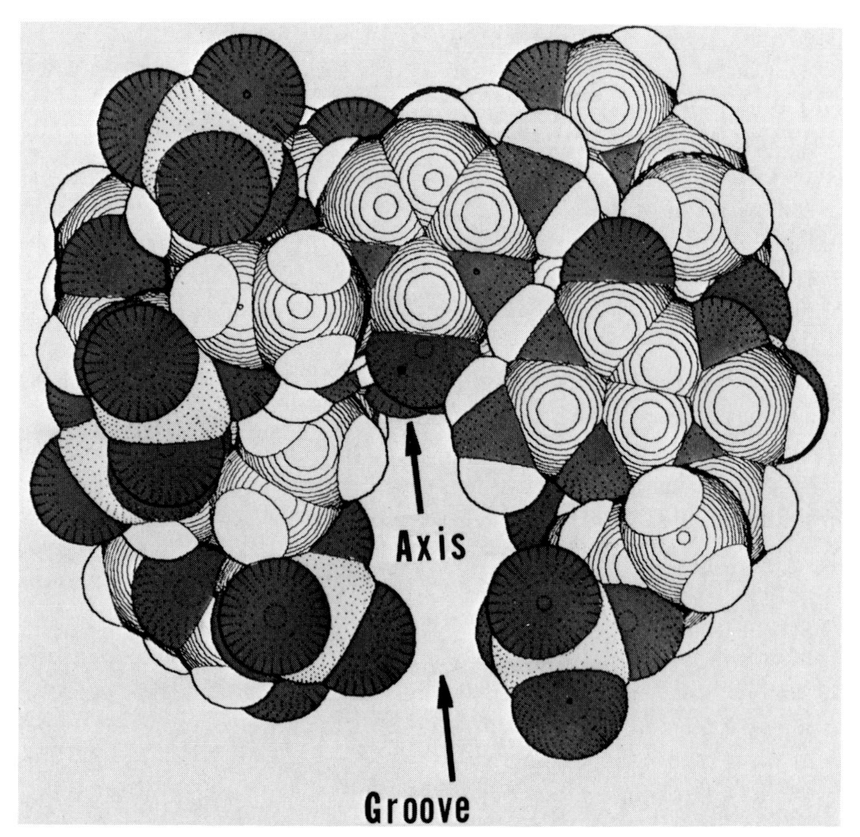

Fig. 5. A van der Waals diagram showing a cross section of Z-DNA as viewed down the helix axis. A 3-bp segment is illustrated. It can be seen that the deep groove extends almost to the helical axis, which is indicated by a solid dot. In this diagram the oxygen and nitrogen atoms are shaded. The phosphorus atoms are stippled and the carbon atoms are drawn with thin concentric circles. Hydrogen atoms are unshaded.

bilize Z-DNA (*4*). The geometry of the AT base pairs in Z-DNA is similar to that of the CG base pairs, with the adenine residues in the *syn* conformation. In this crystal the water molecules in the helical groove of Z-DNA are disordered near the AT base pairs, in contrast to the high level of ordering found in the solvated groove in the segments containing CG base pairs. The local solvent disordering may be related to the fact that AT base pairs form Z-DNA less readily than CG base pairs. It has been known for many years that both purines and pyrimidines can rotate about their glycosyl bonds and these two conformations have been seen in a variety of crystallographic as well as solution studies. However, the Z-DNA structure was the first time that a *syn* conformation was used systematically in the formation of a polynucleotide structure. An early theoretical study (*9*) suggested that although purines could form the *syn* conformation without loss of energy, there was some steric hindrance to the formation of pyrimidine residues in the *syn* conformation. Experimental studies of solutions generally indicate that purine residues can form *syn* conformations relatively easily, but they are less common for pyrimidines (*10*). In Z-DNA, every other residue along the chain is in the *syn* conformation, suggesting that this conformation is more likely to be found in sequences with alternations of purines and pyrimidines. However, although there is some energy loss due to van der Waals crowding when pyrimidines are in the *syn* conformation, the energy loss is not very large.

Other sequences also form Z-DNA. For example, the relative ease of forming a Z-DNA structure with pyrimidines in the *syn* conformation is illustrated with the crystal analysis of d(*CpGpApTp*CpG) where the cytosines have methyl or bromine atoms on their C-5′ positions (*5*). This molecule crystallized as Z-DNA even though two of the six nucleotides were not alternations of purines and pyrimidines (Fig. 6). In this double helix, the two thymine residues are in the *syn* conformation and the adenines are in *anti* conformations. The structure adopts a typical Z-DNA conformation, a left-handed double helix with *syn–anti* conformational alternations of the nucleotides along the helical backbone. The inclusion in Z-DNA of base pairs out of purine–pyrimidine alternation produces changes in the external form of the molecule. This is illustrated in Fig. 6, where the arrows point to the thymidine in the *syn* conformation, which produces a bulge on the outside of the molecule. In addition, the adenine residues in the *anti* conformation lie closer to the axis and a slight external depression is found there. Such changes in external morphology are not found in B-DNA, as the form of that molecule is approximately independent of nucleotide sequence.

IV. Z-DNA in Solution

The synthetic polydeoxynucleotide poly(dG-dC) goes through a cooperative conformational change in solution when the concentration of salt is raised (*11*). In 4 *M* NaCl, the circular dichroism of the solution is nearly inverted compared to the spectrum in 0.1 *M* NaCl. The Z-DNA crystal has a Raman spectrum identical to that of the high-salt form of poly(dG-dC) and quite different from

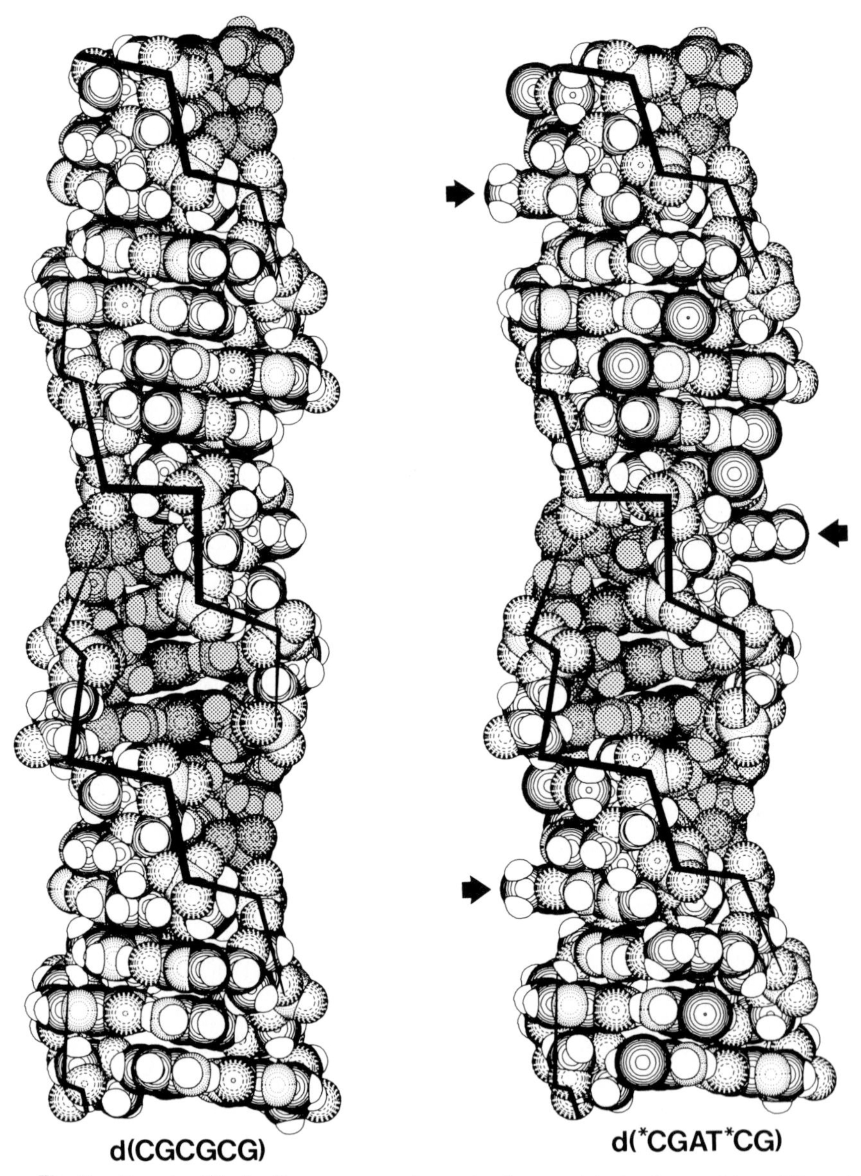

Fig. 6. Van der Waals diagrams are shown for the crystal structures of two different sequences that form Z-DNA. One has an alternating purine–pyrimidine sequence: d (CGCGCG). The other, d(*CGAT*CG), has two base pairs that do not have alternations of purines and pyrimidines. *C has methyl or bromine attached to the C-5 position. The solid

that of the low-salt form (*12, 13*). From this it was inferred that the low-salt form represented right-handed B-DNA, while the high-salt form was identified as Z-DNA.

There is an equilibrium in solution between the right-handed and left-handed forms of DNA. The actual distribution between these two states is strongly influenced by environmental conditions as well as by the sequence of nucleotides. In this equilibrium B-DNA is usually the lower energy state. However, Z-DNA can become the lower energy state when the system is modified in some way to stabilize it. The relative instability of Z-DNA relative to B-DNA is partly associated with the fact that the charged phosphate groups on opposite strands come closer together in Z-DNA than in B-DNA, as seen in Fig. 1. The distance of closest approach of the phosphate groups across the groove in Z-DNA is 7.7 Å, compared to 11.7 Å across the minor groove in B-DNA (*14*). Many factors are known that stabilize or lower the energy of Z-DNA so that the equilibrium shifts in its favor (*4*). Since purines form *syn* conformations more readily than pyrimidines, Z-DNA formation is favored in sequences containing alternations of purines and pyrimidines. Three kinds of regular polymers exist with simple alternations of purine and pyrimidine sequences: poly(dG-dC), poly(dC-dA)·poly(dG-dt), and poly(dA-dT). We can summarize the tendency of pyrimidine-purine dinucleotides in polymers to form Z-DNA as follows: CG > TG = CA > TA (*4*).

V. Z-DNA Is Immunogenic and Forms Specific Anti–Z-DNA Antibodies

Z-DNA can be stabilized in a low-salt solution through chemical bromination (*15, 16*). When poly(dG-dC) is placed in a 4 *M* salt solution, it assumes the Z-DNA conformation. When Br_2 is added, it reacts largely with the C-8 position of guanine and this bromination prevents guanine from adopting the *anti* conformation. When the salt is dialyzed away from brominated poly(dG-dC), it remains as Z-DNA with only one guanine C-8 hydrogen atom in three replaced by bromine. This stabilized form of Z-DNA is a strong immunogen, producing antibodies with a high degree of specificity for Z-DNA (*15*). In contrast, B-DNA is a very

line goes from phosphate to phosphate group. The arrows point to the thymine residues in the *syn* conformation that protrude away from the axis of the molecule.

poor immunogen. Monoclonal antibodies have also been produced against Z-DNA (*17–19*). Some monoclonal antibodies bind to the base pairs on the surface of Z-DNA, while other appear to have a preference for the sugars and negatively charged phosphate groups.

Using the method of indirect immunofluorescence, anti–Z-DNA specific antibodies have been used to look for Z-DNA in cells. In these studies an anti–Z-DNA antibody raised in rabbits, for example, is added to a fixed cytological preparation. A second antibody is then added that was raised in goats against the first antibodies. The goat antibody has a fluorescent chromophore conjugated to it that is visualized by illuminating the preparation at a wavelength that excites fluorescence. Photographs are taken at the emitting wavelength so that position of the initial antibodies can be seen. Several organisms have been studied and fluorescent patterns can be visualized in their genome (*4*).

VI. Negative Supercoiling Unwinds B-DNA and Stabilizes Z-DNA

In general, DNA in biological systems is under topological constraint, which means that the long molecule is twisted about itself or supercoiled. Supercoiling exists whenever the number of turns of the double helix is not equal to the number of turns the molecule would adopt if it were in a linear or relaxed form. *In vivo,* DNA is generally not found in a linear form but is either in a circular form as in plasmids or it is constrained in topological domains in the genome. A complex series of enzymes, the topoisomerases, keep DNA in a negatively supercoiled or slightly underwound state.

In eukaryotic chromatin, negative superhelical turns are largely taken up in the DNA coiling around nucleosomes so that most of the DNA is not torsionally strained even though it is supercoiled. However, it is likely that some of the DNA is subjected to transient torsional stress. This would be especially true if nucleosomes dissociate when the chromatin is transcriptionally active. Negatively supercoiled DNA has a higher free energy than relaxed DNA. The free energy of supercoiling is proportional to the square of the number of superhelical turns or the superhelical density. The free energy of supercoiling can be used to change DNA, since processes that reduce the number of superhelical turns are energetically favored. In negatively supercoiled DNA, these processes include unpairing of bases with strand separation and unwinding of the double helix, as well as binding to proteins as in nucleosome formation. Another process that is facilitated in a negatively supercoiled plasmid is stabilization of Z-DNA (*20, 21*).

The manner in which negative supercoiling stabilizes Z-DNA is shown in Fig. 7. There we can see the effect on supercoiling of converting one turn of right-handed B-DNA into a left-handed Z-DNA segment. The negative supercoiling energy in the negative supercoils is adequate to stabilize one turn of left-handed Z-DNA. This effect can be visualized in a number of different experimental modes. One of the simplest is a nitrocellulose filter binding experiment. Naked DNA passes through holes in the filter, but the DNA is retained if protein is bound to it. When a plasmid is supercoiled sufficiently to induce the formation of Z-DNA, it is detected by retention of the plasmid–antibody complex on the filter (*22*). Relaxed plasmids are not retained on the filter in the presence of the antibody. At a critical superhelical density, the plasmids begin to be retained. As the number of negative superhelical turns increase there is a corresponding increase in the amount of retention. This demonstrates the manner in which increasing negative superhelical densities result in the induction of Z-DNA formation as detected by the binding of anti-Z-DNA antibodies. The complex of antibodies bound to negatively supercoiled plasmids can also be visualized in the electron microscope (*22*).

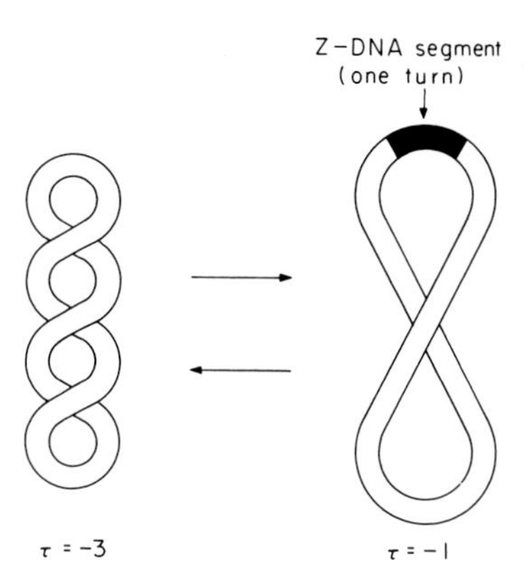

Fig. 7. A schematic diagram in which the double helix is represented as a tubular structure. The diagram on the left represents a negatively supercoiled plasmid with three negative supercoiled turns. In the diagram at the right, a 12-bp segment of B-DNA has converted to Z-DNA, and this results in a loss of two superhelical turns, from $\tau = -3$ to $\tau = -1$. The energy of supercoiling is used to stabilize Z-DNA.

It is possible to map the position of Z-DNA formation within a supercoiled plasmid using the antibody binding sites. Glutaraldehyde cross-linking is used to fix the antibody covalently on the DNA. This is followed by restriction endonucleolytic DNA fragmentation and nitrocellulose filtration. DNA fragments containing the cross-linked antibody will be retained on the filter and will thus be missing from the filtrate, which is analyzed by gel electrophoresis. When this experiment is carried out with the plasmid pBR322, a 14-bp segment of alternating purines and pyrimidines with one base pair out of alternation is found to form Z-DNA (*22*). Further evidence for the formation of Z-DNA in this region of plasmid pBR322 is shown by the fact that the cleavage of three restriction endonucleases encoded in this segment (*Hae*III, *Hha*I, and *Sau*3A) are also inhibited by the presence of the antibody (*23*). This is a technique that can be used for locating segments of Z-DNA in any plasmid.

Such experiments demonstrate that negative supercoiling is a powerful factor in Z-DNA stabilization. However, it should be emphasized that the actual formation of Z-DNA *in vivo* will be strongly influenced by the presence of Z-DNA binding proteins that also stabilize Z-DNA to a considerable extent. Furthermore, they may significantly alter the distribution of Z-DNA which one would otherwise observe in the absence of the protein. It is important to understand the interaction of negative supercoiling that generates torsional strain in DNA and its relation to Z-DNA binding proteins. Z-DNA binding proteins may relieve torsional strain in DNA upon binding and that drives the reaction.

VII. In Chromatin, Z-DNA is Stabilized by Specific Protein Binding

Proteins are the major macromolecules with which the nucleic acids interact. A class of proteins exists that binds specifically to left-handed Z-DNA but not to B-DNA. These proteins have been found in the nuclei of eukaryotic cells. Z-DNA binding proteins were initially isolated from the nuclei of *Drosophila* cells using the method of affinity chromatography (*24*). Brominated poly(dG-dC), stable in the Z form at low NaCl concentrations, was attached covalently to Sephadex G-25. *Drosophila* nuclei were lysed and their proteins were dissociated from DNA by salt. After removal of the DNA, B-DNA binding proteins were precipitated and the remaining proteins were absorbed to the Z-DNA affinity column. Proteins were eluted from the column with salt and were assayed for binding to [^{3}H]Br-poly(dG-dC) (Z-DNA) or [^{3}H]poly(dG-dC) (B-DNA), using nitrocellulose filter binding. Polyacrylamide gel analysis of these Z-DNA binding proteins revealed five major bands and several minor bands. Some of the major bands could also be seen in a gel analysis of total unfractionated proteins from the *Drosophila* cell nucleus. The major bands migrated with apparent molecular weights greater than 70,000. It appears that there are numerous species of Z-DNA binding proteins in the *Drosophila* nucleus, and some of them seem to be moderately abundant.

Using DNA polymers, filter binding assays showed that the *Drosophila* Z-DNA binding proteins can shift the equilibrium from the B to the Z conformation and hold it there (*24*).

In general, one would anticipate that Z-DNA binding proteins are likely to be of two different types. Those that recognize Z-DNA by binding to its backbone would be relatively insensitive to sequence. A second type of protein would bind to the bases as well, and these are likely to be sensitive to nucleotide sequence also.

Another tissue that has been studied extensively is wheat germ (*25*). Using techniques similar to those that were employed in the isolation of *Drosophila* Z-DNA binding proteins, a series of Z-DNA binding proteins has been isolated with molecular weights that vary from 40,000 to nearly 150,000. Like the *Drosophila* proteins, these have the ability to bind to brominated poly(dG-dC) (Z-DNA) but not to poly(dG-dC) (B-DNA). It can also be shown that the antibody against Z-DNA competes with the binding of wheat germ Z-DNA binding proteins. This reinforces the idea that they are binding at the same Z-DNA site. Filter binding experiments were also carried out with negatively supercoiled plasmids containing inserts such as $d(CG/GC)_n$ or $d(CA/GT)_n$ in which n varies from 10 to 60. The specificity of Z-binding proteins can be tested in systems containing plasmids with or without such a Z-DNA–forming insert. Thus, when the two plasmids are supercoiled to the same negative superhelical density, there is usually a much greater retention on the filter paper of the plasmid with the Z-DNA–forming insert than the plasmid without the insert. A test of this type clearly demonstrates that the protein binding is due to Z-DNA and is not, for example, due to proteins that bind to supercoiled DNA *per se*.

In the experiments cited above with both *Drosophila* and wheat germ, the protein preparations were found to stabilize Z-DNA formation in inserts of $d(CA/GT)_n$ to a greater extent than inserts with $d(CG/GC)_n$. This may reflect the fact that eukaryotic DNA contains many segments of $d(CA/GT)_n$, in which $n >$ 25 (*26*), and it is possible that these proteins stabilize those segments.

It is likely that some proteins can stabilize specific sequences of DNA. In an attempt to find such proteins, experiments were carried out on the minichromosome formed by the tumor virus SV40 (*27*). The tumor virus SV40 infects monkey kidney cells and transforms the cells. The virus goes through an active transcription phase, and it can be isolated as a small fragment of chromatin containing 24 nucleosomes associated with each covalently closed double helical SV40 genome containing slightly over 5 kb of DNA. One can ask whether there are specific proteins associated with the minichromosome that binds Z-DNA. To answer this question,the SV40 minichromosome was incubated with either labelled Z-DNA or B-DNA at varying salt concentrations. Salt is known to dissociate protein–nucleic acid interactions and its addition would solubilize proteins that could then bind either to Z-DNA or B-DNA in the medium. The results of an experiment of this type is illustrated in Fig. 8. Radioactive Z-DNA or B-DNA was added to a solution containing SV40 minichromosomes incubated at varying salt concentrations. After incubation the radioactive DNA was passed through a nitrocellulose filter. It can be seen that there is very little release of protein that binds to B-DNA until the salt concentration is raised above 0.6 *M* NaCl. At these higher salt concentrations, histones are liberated and retention of B-DNA is observed. When the experiment is repeated using radioactive Z-DNA, a different result is seen. At low salt, near 0.2 *M* NaCl, there is a significant solubilization of proteins that have the capacity to bind to Z-DNA. The retention of Z-DNA is seen to remain virtually constant until the salt concentration is raised above 0.6 *M* NaCl, at which point histones are liberated from chromatin. The experiment in

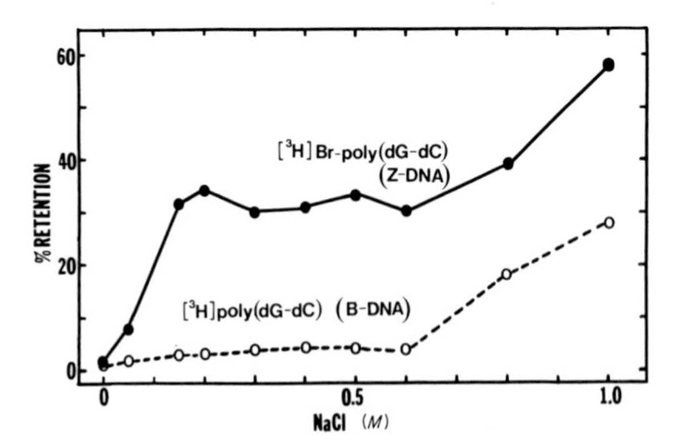

Fig. 8. A diagram showing the DNA binding properties of proteins released from SV40 minichromosomes as a function of increasing salt concentration. Radioactive B- or Z-DNA is added to the incubation mixture and the liberation of B- or Z-DNA binding proteins is measured by the extent to which these radioactive DNAs are retained on nitrocellulose filtration.

Fig. 8 suggests that there is a difference in conditions under which Z-DNA binding proteins are solubilized contrasted to B-DNA binding proteins. Z-DNA binding proteins appear to be less tightly bound, as judged by the ease with which they are released from the SV40 minichromosome at lower salt concentrations. Although the curves in Fig. 8 were obtained with the SV40 minichromosome, very similar curves are obtained from chromatin in general from a variety of sources (F. Azorin and A. Rich, unpublished data).

The proteins solubilized in 0.2 *M* salt from the SV40 minichromosome were isolated by incubation with Z-DNA covalently bound to Sephadex. A low-speed centrifugation made it possible to sediment the Sephadex Z-DNA complex and its attached Z-DNA binding proteins. These were then analyzed on polyacrylamide gels and compared to both the native minichromosome and the minichromosome from which Z-DNA binding proteins have been removed. This analysis showed a number of discrete bands that had Z-DNA binding activity, the most prominent of which was a very large protein with a molecular weight near 200,000. The proteins that were liberated in this manner were called the PZ extract, indicating that they were solubilized proteins that bound to Z-DNA (*27*).

VIII. Negatively Supercoiled SV40 DNA Forms Z-DNA

By using anti–Z-DNA antibodies cross-linked to DNA, we can search any circular genome for segments that can form Z-DNA. The simian virus 40 (SV40), a DNA tumor virus, has 5243 bp and it codes for early and late proteins. The transcription of these proteins is regulated by a control region near the origin of replication. The control region is only a few hundred nucleotides long, but it carries out a number of functions including regulation of viral replication and transcription. It also contains a transcriptional enhancer, that is, a small segment of DNA which by its presence insures a high level of transcriptional activity for the early transcriptional unit. In SV40, the transcriptional enhancer is associated with a region that contains two 72-bp repeated segments (*28, 29*).

In infected host cell nuclei, the SV40 genome is found in a minichromosome, a nucleosomal chromatin structure. The nucleosomes on the minichromosomes are not distributed at random over the DNA, but a nucleosomal free gap is found over the control region near the origin of replication. Although negative superhelical turns are absorbed by the nucleosomes of the minichromosome, there has been a report of torsional strain in transcriptionally active SV40 minichromosomes (*30*).

Experiments were carried out to see whether Z-DNA forms in negatively supercoiled SV40 DNA. The negatively supercoiled SV40 was cross-linked to anti–Z-DNA antibodies and a restriction fragment was retained on the nitrocellulose filter. This included the control region spanning the origin of replication through to the transcriptional enhancer, some 273 bp (*31*). Although the plasmid contains over 5 kb of DNA, Z-DNA formation occurs only in the regulatory region of the virus. Additional experiments were carried out to locate more precisely the Z-DNA forming regions and they pointed to segments in and near the 72-bp repeated sequences. This includes a region near the sites for the restriction enzyme *Sph*I, but no Z-DNA formation was seen 130 bp away at the origin of replication near the *Bgl*I restriction site.

IX. Interaction of Z-DNA Binding Proteins with SV40 DNA

Incubation of SV40 minichromosomes in a 0.2 *M* salt solution liberates Z-DNA binding proteins that can be isolated through their ability to combine to Z-DNA linked to Sephadex. The DNA binding properties of this protein mixture, called a PZ extract, was studied. As might be expected, it binds readily to brominated poly(dG-dC) (Z-DNA) but not to poly(dG-dC) (B-DNA). However, it also binds to Z-DNA that is stabilized in negatively supercoiled DNA circles. The results of such an experiment are shown in Fig. 9, where the binding of the PZ extract is shown both to supercoiled and relaxed plasmid pBR322 or SV40 DNA (*27*). Both of these circular DNAs are known to form Z-DNA, as judged by their ability to bind the Z-DNA antibodies. Here it can be seen that gradually increasing the amount of the PZ extract proteins leads to retention of the supercoiled circular DNAs but not to retention of the relaxed DNAs. Both DNAs were supercoiled to the same negative superhelical density, but a somewhat greater retention was observed with SV40 DNA.

The binding of the PZ extract protein to Z-DNA segments in both plasmids was also demonstrated through their ability to compete with the binding antibodies against Z-DNA. In both cases, a significant competition was observed between antibody binding and the proteins of the PZ extract. However, the proteins of the PZ extract bound much more tightly to negatively supercoiled

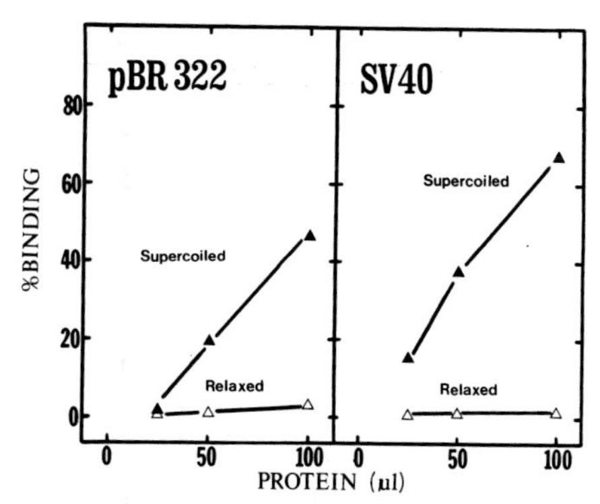

Fig. 9. Filter binding experiments are shown for the plasmid pBR322 and SV40 DNA in the presence of varying amounts of Z-DNA proteins obtained from SV40 minichromosomes. It can be seen that the relaxed DNA is not bound to the filter but supercoiled DNAs are bound. Both supercoiled preparations have a superhelical density of −0.10.

SV40 than to negatively supercoiled pBR322. This is consistent with the idea that there may be a sequence-specific component to the Z-DNA binding observed with negatively supercoiled SV40 that is not present in the plasmid pBR322.

X. Discrete Location for Z-DNA Binding Proteins on SV40 DNA

To ascertain whether the Z-DNA binding proteins were occupying a specific site on the SV40 genome, restriction endonuclease experiments were carried out. The first of these experiments involved naked negatively supercoiled SV40 DNA and the rate of restriction endonuclease cleavage was measured both in the presence and absence of Z-DNA binding proteins (*27*). Three restriction endonucleases were used. One of these was *Sph*I, which cleaves in the 72-bp repeat region of the SV40 genome [this is part of the transcriptional enhancer that was identified as a site forming Z-DNA (*31*)]. Another was the restriction endonuclease *Bgl*I, which is approximately 130 bp away from an *Sph*I site and is at the origin of replication. A third endonuclease is *Eco* RI, which is far removed from the control region. The presence of Z-DNA binding proteins did not materially influence the rate of cleavage of supercoiled SV40 DNA by either *Eco*RI or *Bgl*I but it did seriously retard the rate of cleavage by *Sph*I. This suggested that the Z-DNA binding proteins may be lying at or near the *Sph*I site.

To test this further, experiments were then carried out with the SV40 minichromosome. In these experiments, the digestion of SV40 DNA was measured in the intact minichromosome as well as with SV40 minichromosomes that had Z-DNA binding proteins removed. The results of these experiments are shown in Fig. 10. The presence or absence of Z-DNA binding proteins has no effect on the cleavage of SV40 DNA in minichromosomes by either *Eco*RI or *Bgl*I. However, removal of the Z-DNA binding proteins has a significant effect in opening up sites for cleavage by *Sph*I. There is a doubling of the cleavage associated with removal of the Z-DNA binding proteins. The effect shown in Fig. 10 is also reversible. SV40 minichromosomes were prepared and the PZ proteins were removed from them. These preparations with an enhanced sensitivity to *Sph*I cleavage then had the PZ extract added back to it. This results in a reversal of the enhanced cleavage shown in Fig. 10. The Z-DNA binding proteins can thus be removed, and the *Sph*I site is exposed. The proteins can then be returned to the minichromosome and these sites no longer have the enhanced susceptibility to cleavage. It is interesting that no alteration is found in the *Bgl*I site even though it is only 130 nucleotides away.

These experiments suggest that a Z-DNA binding protein is sitting on or near the *Sph*I site in the SV40 minichromosome. Not all of the minichromosomes have this protein, since only ~25% have enhanced cleavage, as shown in Fig. 10. For some time it has been known that 15–25% of the SV40 minichromosomes is transcriptionally active. It is possible that the subfraction that we are looking at in Fig. 10 may be a group that has a specialized physiological state such as being transcriptionally active. Further experimentation will have to be carried out in order to establish this point.

The significant result from these experiments is that a Z-DNA binding protein is found that appears to be positioned near the transcriptional enhancer of SV40 in the control region. Finding a protein in that position strongly suggests that it may participate in some way in regulating the various functions controlled by this segment of DNA in the virus. This could include a transcription enhancer activity or another role in transcription that is yet to be determined.

At the present time, we see a large number of Z-DNA binding proteins found generally in the nuclei of cells. In most cases, we do not know the physiological role of these proteins except that they are capable of stabilizing the higher energy left-handed form of the helix. One may imagine a number of different mechanisms in which these proteins could be engaged. For example, removal of a Z-DNA binding protein by some agent such as RNA polymerase would result in the

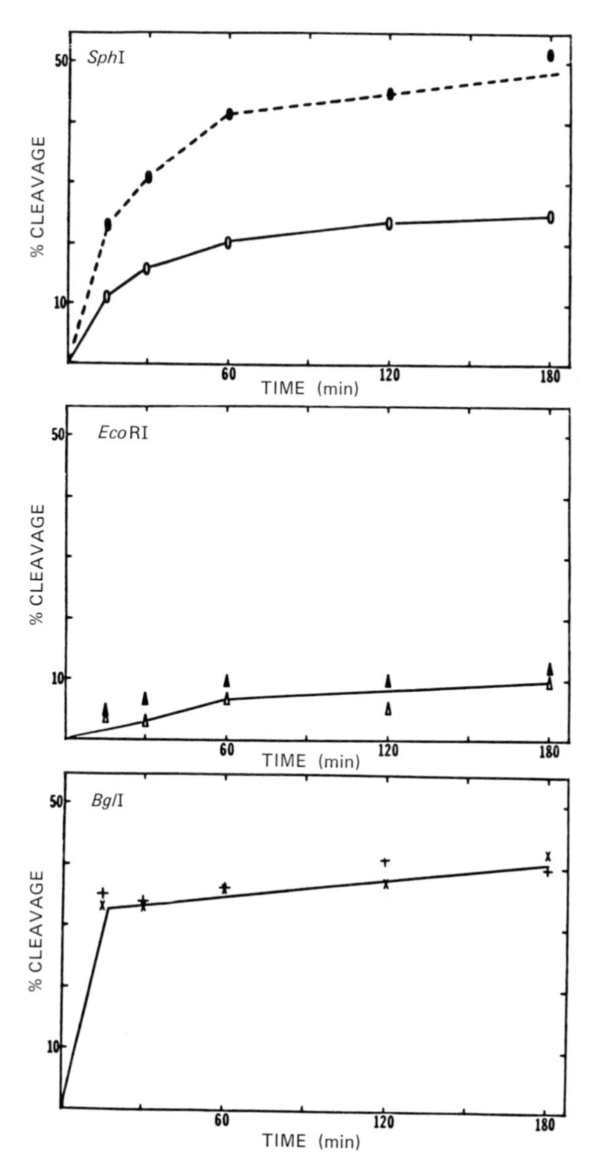

Fig. 10. The accessibility of restriction endonuclease cleavage sites on SV40 minichromosomes or PZ binding protein–depleted SV40 minichromosomes is shown *(27)*. Following incubation of the minichromosome with the restriction endonuclease, the DNA was isolated and an agarose gel used to separate the cleaved DNA from the intact DNA. This was done for various time periods. Native SV40 minichromosomes are shown by the symbols ○, △, +. The PZ depleted SV40 minichromosomes are shown by the symbols ●, ▲, ×. Only the *Sph*I site is rendered accessible by removal of the PZ binding proteins.

conversion of Z-DNA into B-DNA. This conversion event could be used in a number of ways. It might, for example, facilitate the attachment of RNA polymerase to the DNA. Since that event may require separation of the two DNA strands, such strand separation may occur at the B–Z junction. Alternatively, it would be facilitated once B-DNA is formed, since there is a large input of negative superhelical energy associated with the conversion from Z-DNA to B-DNA. Superhelical energy facilitates strand separation. Experiments to be carried out in the near future will be oriented toward elucidating such detailed mechanisms so that we can more specifically define the role of Z-DNA.

Acknowledgments

This research was supported by grants from the National Institutes of Health, the American Cancer Society, the National Aeronautics and Space Administration, the Office of Naval Research, and the National Science Foundation.

References

1. J. D. Watson and F. H. C. Crick, *Nature (London)* **171,** 737 (1953).
2. R. E. Franklin and R. Gosling. *Nature (London)* **171,** 740–742 (1953).
3. A. H.-J. Wang, G. J. Quigley, F. J. Kolpak, J. L. Crawford, J. H. van Boom, G. van der Marel, and A. Rich, *Nature (London)* **282,** 680–686 (1979).
4. A. Rich, A. Nordheim, and A. H.-J. Wang, *Ann. Rev. Biochem.* **53,** 791–846 (1984).
5. A. H.-J. Wang, R. Gessner, G. van der Marel, J. H. van Boom, and A. Rich, *Proc. Natl. Acad. Sci. U.S.A.* **82,** 3611–3615 (1985).
6. J. M. Rosenberg, N. C. Seeman, J. J. P. Kim, F. L. Suddath, H. B. Nicholes, and A. Rich, *Nature (London)* **243,** 150–154 (1973).
7. R. O. Day, N. C. Seeman, J. M. Rosenberg, and A. Rich, *Proc. Natl. Acad. Sci. U.S.A.* **70,** 849–853 (1973).
8. A. H.-J. Wang, T. Hakoshima, G. van der Marel, J. H. van Boom, and A. Rich, *Cell* **37,** 321–331 (1984).
9. A. E. V. Haschemeyer and A. Rich, *J. Mol. Biol.* **27,** 369–384 (1967).
10. D. B. Davies, *in* "Progress in NMR Spectroscopy" Vol. 12, p. 135–186. Pergamon, London, 1978.
11. F. M. Pohl and T. M. Jovin, *J. Mol. Biol.* **67,** 375–396 (1972).
12. F. M. Pohl, A. Ranade, and M. Stockburger, (1973) *Biochim. Biophys. Acta* **335,** 85–92 (1973).
13. T. J. Thamann, R. C. Lord, A. H.-J. Wang, and A. Rich, *Nucleic Acids Res.* **9,** 5443–5457 (1981).
14. A. H.-J. Wang, G. J. Quigley, F. J. Kolpak, G. van der Marel, J. H. van Boom, and A. Rich, *Science* **211,** 171–176 (1981).

15. E. M. Lafer, A. Moller, A. Nordheim, B. D. Stollar, and A. Rich, *Proc. Natl. Acad. Sci. U.S.A.* **78,** 3546–3550 (1981).
16. A. Moller, A. Nordheim, S. A. Kozlowski, D. Patel, and A. Rich, *Biochemistry* **23,** 54–62 (1983).
17. A. Moller, J. E. Gabriels, E. M. Lafer, A. Nordheim, A. Rich, and B. D. Stollar, *J. Biol. Chem.* **257,** 12081–12085 (1982).
18. F. M. Pohl, *Cold Spring Harbor Symp. Quant. Biol.* **47,** 113–118 (1983).
19. R. Thomae, S. Beck, and F. M. Pohl, *Proc. Natl. Acad. Sci. U.S.A.* **80,** 5550–5553 (1983).
20. C. K. Singleton, J. Klysik, S. M. Stirdivant, and R. D. Wells, *Nature (London)* **299,** 312–316 (1982).
21. L. J. Peck, A. Nordheim, A. Rich, J. C. Wang, *Proc. Natl. Acad. Sci. U.S.A.* **79,** 4560–4564 (1982).
22. A. Nordheim, E. M. Lafer, L. J. Peck, J. C. Wang, B. D. Stollar, and A. Rich, *Cell* **31,** 309–318 (1982).
23. F. Azorin, A. Nordheim, and A. Rich, *EMBOJ.* **2,** 649–655 (1983).
24. A. Nordheim, P. Tesser, F. Azorin, Y. H. Kwon, A. Moller, and A. Rich *Proc. Natl. Acad. Sci. U.S.A.* **79,** 7729–7733 (1982).
25. E. Lafer, R. Sousa, B. Rosen, A. Hsu, and A. Rich, *Biochemistry* (in press), (1985).
26. H. Hamada, M. G. Petrino, and T. Kakunaga, *Proc. Natl. Acad. Sci. U.S.A.* **79,** 6465–6469 (1982).
27. F. Azorin and A. Rich, *Cell* **41,** 365–374 (1985).
28. C. Benoist and P. Chambon, *Nature (London)* **290,** 304–310 (1981).
29. P. Gruss, R. Dhar, and G. Khoury, *Proc. Natl. Acad. Sci. U.S.A.* **78,** 943–947 (1981).
30. A. N. Luchnik, V. V. Bakayev, I. B. Zbarsky, and G. P. Georgiev, *EMBO J.* **1,** 1353–1359 (1982).
31. A. Nordheim, and A. Rich, *Nature (London)* **303,** 674–679 (1983).

Factors Involved in Elongation and Termination of Bacterial and Mammalian Transcription*

MICHAEL J. CHAMBERLIN, J.-F. BRIAT, RUSSELL L. DEDRICK, MICHELLE HANNA, CAROLINE M. KANE, JUDITH LEVIN, REBECCA REYNOLDS, AND MARTIN SCHMIDT

Department of Biochemistry
University of California
Berkeley, California

There is no doubt that interesting and even ingenious experiments have been, and will continue to be, performed with intact cells. Nevertheless, it has been painfully obvious for some time that the distance between the data and the deductions derived from the analyses of intact cells is

*This research was supported by a research grant (GM12010) from the National Institute of General Medical Sciences, and by a fellowship (J-47-83) from The California Division, American Cancer Society.

far too great for certainty. As between the two, many of us would, I suspect, prefer to be ingenious rather than courageous. It is therefore with reluctance that one accepts the conclusion that ingenuity alone will no longer suffice and that the age of courage and the direct approach has arrived.

Spiegelman (1957)

I. Introduction

Transcription is one of the initial steps in gene expression and also plays a pivotal role in the control of cell growth and the initiation of DNA replication in many cells. For this reason, regulation of transcription is of central importance in normal cell growth and in development of both bacterial and mammalian cells. A discussion of transcription regulation is also appropriate for a book on current progress in cancer research; several examples are known of cellular transformation in which activation of transcription by genomic rearrangements or by viral insertions appears to be a primary event in transformation. These are examined in some detail in other chapters.

Over the past 8 years impressive progress has been made in dissecting the structure of eukaryotic genes and mapping sequences that affect their expression. However, relatively little is known about the molecular events that are involved in control of transcription in mammalian cells. As anticipated all too clearly by Spiegelman (*1*) almost 30 years ago, our progress has been limited by our ability to obtain cell-free systems in which transcription occurs with a specificity and efficiency characteristic of the cellular process. Such systems are essential if one is to identify the components involved in transcription and unravel the actual molecular reactions that they control.

There has been promising progress in obtaining systems that allow correct initiation of mammalian transcription, and this has permitted the identification and purification of several factors that may participate in this process (*2–4*). However, transcription is also regulated at the steps of elongation and termination of RNA chains (*5–8*). Current cell-free mammalian transcription systems show important defects in chain elongation, and almost nothing is known about termination events in synthesis of mammalian mRNAs (*9–11*).

We review here recent progress in studies of bacterial transcription that show that RNA chain elongation and termination are complex processes involving a number of factors in addition to the RNA polymerase. Using techniques and approaches drawn from these prokaryotic studies, we have begun to develop systems for the biochemical study of RNA chain elongation and termination by mammalian RNA polymerase II that may facilitate the dissection of these complex processes.

The basic catalytic element in transcription, RNA polymerase, was first identified almost 25 years ago (*12–14*). Since that time studies with isolated DNA and RNA polymerases have provided a good understanding of the basic chemical reactions involved in transcription (*6, 15*). Furthermore, isolated bacterial RNA polymerase holoenzymes can carry out selective initiation and termination of RNA chains for some transcription units in the absence of any additional factors (*16*). This observation led to the long-standing notion that the subunits of the purified RNA polymerase holoenzyme were the only general components of the transcription machinery, with adjunct factors needed for efficient initiation and termination of transcription only at certain promoter and terminator sites.

Over the past 5 years this picture has been called into serious doubt. Biochemical studies show that transcription elongation with purified RNA polymerases is much slower than the cellular process even under optimal reaction conditions (*17*) and that RNA polymerase is not able to traverse certain genes in the absence of additional protein factors (*18, 19*). Such elongation is dominated by ''pausing'' of the polymerase at certain DNA sequences for extended periods (*17, 20*) that are not compatible with the rates of transcription *in vivo*. Genetic studies have identified mutants that lead to rapid shutoff of all bacterial RNA synthesis that do not map in genes for known RNA polymerase subunits (*21, 22*). Finally, studies of transcription during growth of temperate phages such as λ have led to identification of at least three bacterial proteins that appear to be components of an RNA polymerase–antitermination complex (*23, 24*). These proteins are coded for by the *nus* genes, *nusA, nusB,* and *nusE.* The properties of *nusA* and *nusB* mutants suggest that they carry out functions that are essential for the growth of normal bacterial cells as well (*25, 26*). These diverse lines of evidence lead to the conclusion that there are additional protein factors which are probably needed for efficient transcription of all bacterial transcription units. At least some of these factors are probably transcription elongation factors that are associated with the core RNA polymerase during the elongation and termination phases of transcription (*27*).

Isolated mammalian RNA polymerases are also defective in the specificity and efficiency of their action *in vitro.* It has long been known that none of the three different nuclear polymerases acting alone is able to initiate chains at correct promoter sites on DNA *in vitro* (*28*). This finding has led to identification and fractionation of several accessory factors that are essential for correct initiation *in vitro* (*2–4*). However, even in the presence of these factors, transcription by purified RNA polymerase I and II is highly inefficient, and only a small fraction of the RNA polymerase molecules in current preparations can initiate selectively. In addition, the elongation rates found in all *in vitro* systems are much too slow (*29–31*) even in isolated cell nuclei (*32, 33*), and transcription elongation in isolated nuclei stops after proceeding only a few hundred bases (*32, 34–35*). These results suggest that, as in the case of the bacterial transcription machinery, important protein factors needed for normal chain elongation are missing from the purified mammalian RNA polymerase and are even deficient or inactive in isolated nuclei. These factors may play central roles in the transcription of chromatin, which is the true template for gene expression in the mammalian cell.

II. Rationale

Although it appears likely that RNA chain elongation in bacterial cells is carried out by a complex of proteins added to the core RNA polymerase, we do not know the role these components play in transcription nor the identity of all the components involved. We have taken three lines of approach to the study of this problem. First, we have examined the process of RNA chain elongation as it is carried out by purified RNA polymerase acting alone to try to understand what aspects of the reaction are defective and how specific DNA template sequences can control elongation and pausing. These studies have also provided valuable techniques for the assay of basic features of the elongation and termination process. Second, we have developed procedures for the large-scale purification of the *E. coli nusA* and *nusB* proteins, to allow study of their interactions with RNA polymerase and their effects on *in vitro* transcription. Third, we have developed procedures for preparation of a soluble *E. coli* extract which is essentially free from chromosomal DNA fragments. Transcription in this system is completely dependent on added DNA templates. Transcription elongation and termination in this system resemble the processes *in vivo* in many respects, suggesting that missing transcription factors are active and can be identified in such a system.

Using the techniques and concepts drawn from our study of the bacterial transcription elongation–termination system, we have also begun to study elongation and termination along specific DNA sequences using purified mammalian RNA polymerase II. These studies use a highly efficient method for allowing the purified polymerase to initiate at specific template sites and have revealed an unexpected complexity to the elongation reaction carried out by this polymerase.

III. Results and Discussion

A. Elongation Catalyzed by Purified Bacterial RNA Polymerases Is Discontinuous and Is Controlled by RNA Secondary Structure

Some of the earliest studies on the sequencing of RNAs synthesized *in vitro* by bacterial RNA polymerases revealed that chain elongation was discontinuous, with the enzyme pausing at specific sites for extended periods (*36*). When methods became available to follow synchronized transcription through single transcription units (*17, 20*), it became clear that this pausing occurs even under optimal *in vitro* reaction conditions and is found with all DNA templates. It is these pause sites that appear to reduce the rates of *in vitro* transcription well below those of the *in vivo* process.

How do DNA template sequences interrupt the growth of an RNA chain? A clue came from studies on the action of RNA termination sequences (for reviews, see Refs. *5–8*). These sequences (terminators) lead to release of the nascent transcript and the polymerase from the template. At some terminators (often called rho independent), this release can occur spontaneously; at other terminators (rho dependent) the participation of a protein factor, rho, is required. It has been shown that at rho-dependent terminators, the RNA polymerase pauses in the absence of rho factor (*37–39*) and then continues elongation through the terminator region. This has led to a model for RNA chain termination which involves two primary steps, (1) pausing of the polymerase at the terminator site, followed by (2) release of the transcript and polymerase (*6*). At rho-dependent terminators this latter step is catalyzed by rho; at rho-independent sites release is spontaneous, possibly because of the instability of the DNA–RNA hybrid bonds holding the RNA to its template (*40, 41*).

The initial pausing of the polymerase at a terminator appears to be determined by the formation of a specific stem–loop or hairpin structure in the nascent RNA chain at a distance of about ten bases from the growing point of the chain (*5–8,*

42). Most bacterial terminators contain a self-complementary DNA sequence just upstream from the termination site, and mutations that interrupt this complementarity reduce or eliminate termination. Furthermore, the critical secondary structure must form in the RNA, not the DNA strands, since such mutations are only effective when present in the transcribed DNA strand (*43*). An artist's conception of the structure of the ternary RNA polymerase complex during normal elongation, and at a termination site, is shown in Fig. 1. Note that, at this time, there is no direct information as to how the formation of an RNA hairpin blocks further elongation by the polymerase, although possible models have been discussed (*6, 44*).

Our understanding of terminator action leads immediately to a plausible expla-

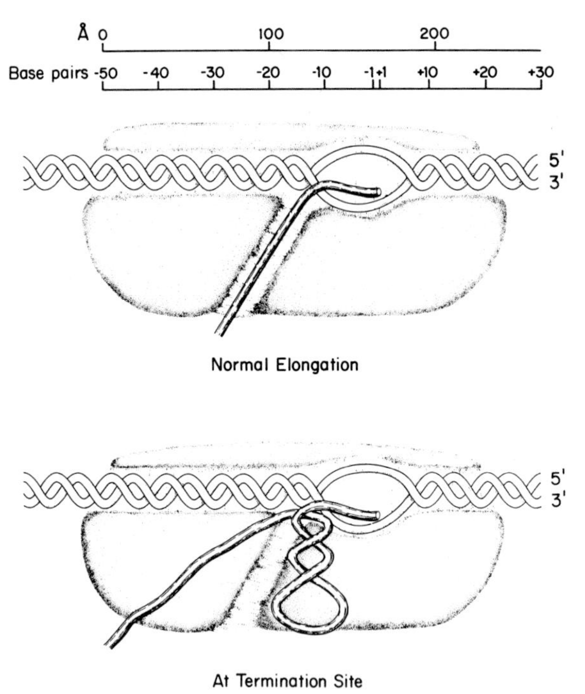

Fig. 1. Diagramatic representation of *E. coli* RNA polymerase during normal chain elongation and at a termination region. Scale shown is that estimated from relative sizes of individual macromolecules; see text and Ref. *6* for review. Although the core RNA polymerase is shown as making contact with almost 70 bp of DNA, our best current information suggests that only about 25 bp of DNA is protected in deoxyribonuclease I footprinting experiments (J. Levin, unpublished studies).

nation as to why RNA polymerase pauses at specific points during elongation: as transcription proceeds, the nascent RNA chain folds into the most stable secondary structure; where this generates a stable hairpin, elongation stops. If no release sequence or factor is present, elongation is presently resumed. Thus the secondary structure of the transcript can modulate both elongation and termination of nascent RNA chains.

Recent studies by Platt and by Yanofsky provide direct support for this model. Introduction of a self-complementary region into a transcript leads to a pause after that site (*45*), and mutations that interrupt such self-complementary regions suppress the corresponding pause (*7*). Pauses can also be suppressed by transcription in the presence of an oligonucleotide complementary to one arm of the hairpin region (*46*).

However a number of aspects of transcriptional pausing and termination are left unclear by the simple model proposed above. These include

1. Are all pause sites due to formation of RNA hairpins? If so, what are the structural parameters needed for a hairpin to function?
2. Some pause sites are release sites for rho factor, others are not (*17*). Also, pausing at some sites, but not others, is enhanced by factors such as *nusA* protein (*17*). Is there a structural basis for this heterogeneity?
3. What are the factors that suppress pausing *in vivo* and allow the rapid elongation characteristic of cellular transcription?

To gain more information about these questions we have set out to map and to study a large collection of transcriptional pause sites. With such a collection, structural features that are common to such sites should become apparent and their role can then be tested by site-directed mutagenesis.

The sites we have mapped lie in the early genetic region of phage T7, which provides an ideal biochemical system for such a study (Fig. 2). Transcription of the region is controlled by three strong promoters located at the left end of the genome. Initiation at the first of these (T7A1) occurs efficiently in the presence of only three of the nucleoside triphosphates (ATP, GTP, and CTP) if the dinucleotide ApU is also provided. These conditions lead to synthesis of a 20-nucleotide RNA which has the sequence

5′ ApU CGAGAGGGACACGGCGAA

(*47, 48*). Since the next nucleotide needed is *U,* the polymerase stops elongation at nucleotide A20. These T7 A20 ternary complexes are quite stable and can be

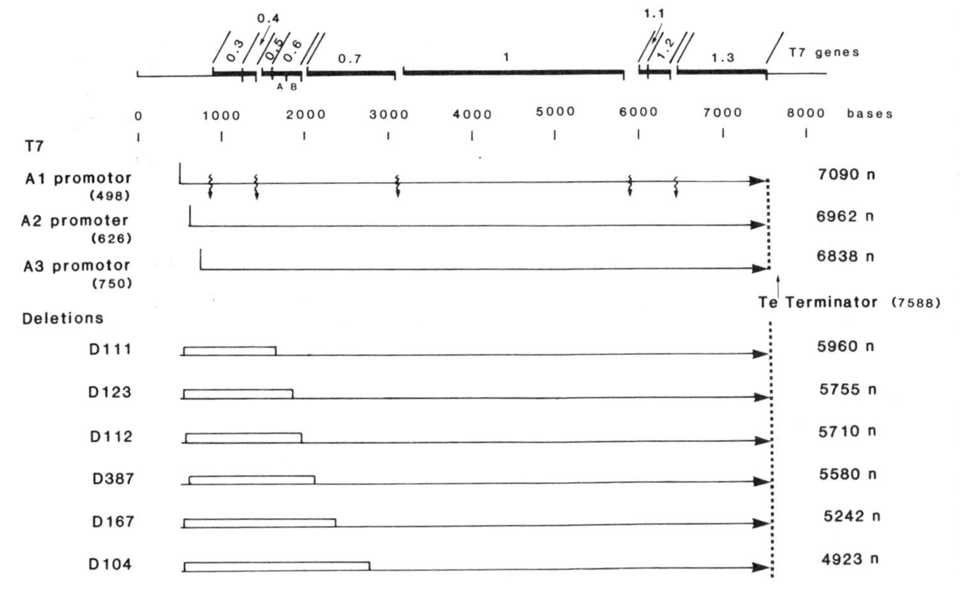

Fig. 2. Transcription and translation signals in the early region of bacteriophage T7 DNA and for several T7 deletion mutants. Maps shown are taken from the DNA sequence of Dunn and Studier (*48*). Distances shown are in nucleotide pairs from the left end of the DNA. Transcript lengths shown on the right of figure are those read from the T7A1 promoter to the early terminator. Wavy arrows mark sites of ribonuclease III processing.

isolated by gel exclusion chromatography. When T7 A20 complexes are added to a mixture of all four triphosphate substrates, synchronous elongation begins immediately and transcription pause sites in the first 150 nucleotides of the transcript can be mapped exactly by analysis of the size transcripts on RNA sequencing gels, along with appropriate markers (Fig. 3).

We can extend our analysis of pause sites through the first 3000 nucleotides of the T7 early region by taking advantage of genetic deletions that remove progressively larger segments beginning just to the right of the T7A1 promoter and extending down to the beginning of the first essential T7 gene, gene *1* (Fig. 2). A map of pause sites in six such deletions is shown in Fig. 4.

The location and strength of these pauses are not significantly changed by small variations in temperature or the ionic conditions used for *in vitro* transcription. However, at very low triphosphate concentrations, the extent of pausing is greatly increased (*17*), and the precise position of any given pause appears to be determined by the relative concentrations of the four nucleoside triphosphates.

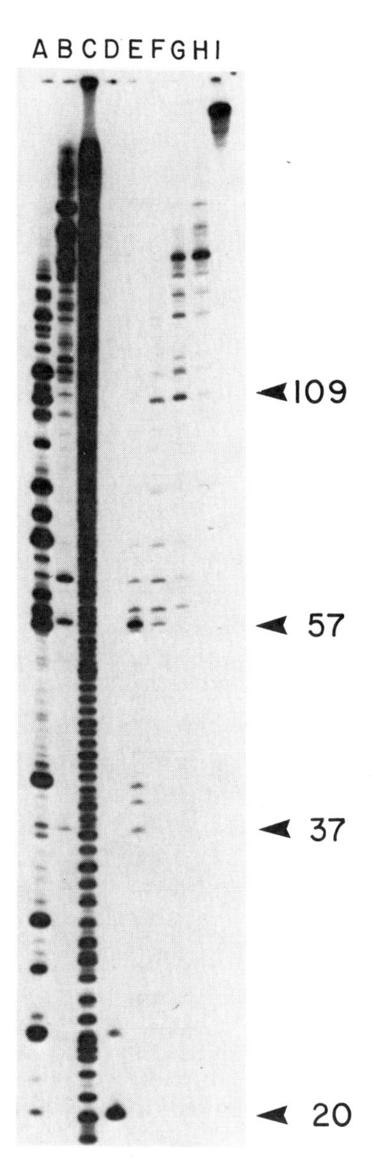

Fig. 3. RNA gel analysis of synchronized RNA chain elongation from T7ΔD111 A20 complexes to map pause sites. T7 A20 complexes were formed by incubation of *E. coli* RNA polymerase (8 μg/ml) with the dinucleotide ApU (50μ*M*), ATP, [α-^{32}P]CTP, GTP (5 μ*M* each), and T7ΔD111 DNA (1.3 m*M*) under standard ionic conditions (*17*), for 9 min at 30°C. Rifampicin (200 μg/ml) was then added to block further initiation, and after an additional 60

(*cont.*)

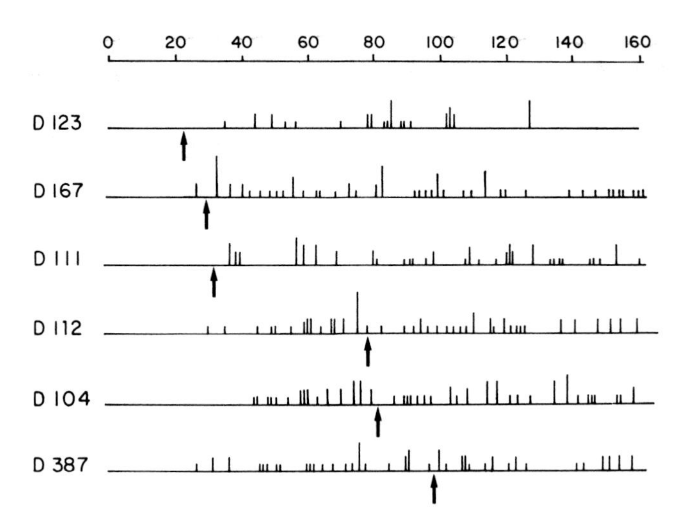

Fig. 4. Map of transcriptional pause sites in promoter–proximal regions of several T7 deletion mutant DNAs. Maps show the location of transcriptional pause sites observed during elongation from T7 A20 complexes under reaction conditions shown in Fig. 3 with six different T7 deletion mutant DNAs as templates. The height of the bar marking each pause is proportional (one, two, or three units) to the strength of the pause as estimated from RNA gels. Arrows below the maps indicate the sites of recombination from which deletions were formed (*47, 48*); see Fig. 2 for maps of deletions.

Analysis of the nucleotide sequences at these different pause sites generally supports the notion that pausing is controlled by the secondary structure of the RNA transcript. We have used a computer analysis program (*49*) to analyze the stability of potential RNA secondary structures expected from studies of model

---

sec incubation, all four nucleoside triphosphates were added to give a final concentration of 10 μ*M* each. Samples were removed after addition of rifampicin (zero time sample) and at various times after addition of all four triphosphates and were analyzed by electrophoresis on a 15% acrylamide–7*M* urea sequencing gel. Individual gel tracks: (A and B) transcription reactions designed to emphasize transcripts paused before U and A residues, respectively, to provide a sequence to index the ladder in track C; (C) ladder produced by partial alkaline hydrolysis of a [γ-^{32}P]ATP-labeled T7ΔD111 A1 transcript; (D–H) samples taken after 0, 1, 2, 3, and 4 min of elongation by T7 A20 complexes; (I) sample taken after 5 min of elongation at 10 μ*M* NTP, followed by 5 min further elongation at 400 μ*M* NTP. Numbers on the right show the length of RNA transcripts at corresponding pause sites in nucleotides from the T7 A1 start site as determined by comparison with the ladder in track C.

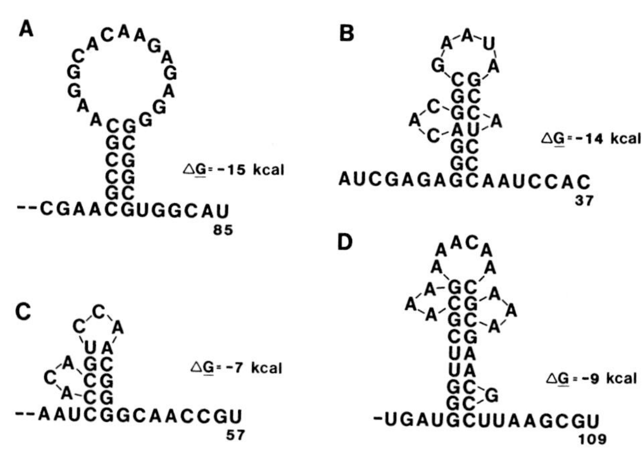

Fig. 5. Plausible secondary structures for RNA transcripts at pause sites seen with T7ΔD111 and T7ΔD123 templates. Structures shown and their relative stabilities were calculated for folding of each full length T7 A1 transcript using the computer analysis from Ref. *49*: (A) the T7ΔD123 transcript at the 85-nucleotide pause site; (B, C, and D) T7ΔD111 transcripts at the 37-, 57-, and 109-nucleotide pause sites, respectively.

structures (*50, 51*). At some pause sites perfect RNA hairpin structures are predicted, analogous to those found at terminators; an example is the pause at nucleotide 85 of T7ΔD123 (Fig. 5A). More commonly there is a stable hairpin structure in which one or more bases are looped out of the hairpin stem; examples are the pause sites at nucleotides 37, 57, and 109 in T7ΔD111 (Figs. 5B, 5C, and 5D). These structures still have predicted free energies that are substantial ($\Delta G = -7$ to -14 kcal/mol).

However, the correlation between RNA secondary structure and RNA polymerase pause sites is broken in two kinds of situations. First, there are RNA hairpins that are predicted to have stabilities in this same range (-7 to -9 kcal/mol) for which there are no corresponding pause sites detected (Figs. 6A and 6B). Second, there are pause sites at which the only potential structures predicted from the rules of Tinoco *et al.* (*50*) are of negligible stability (Fig. 6C).

In the former case, it may be that these potential hairpins do not form because of interference due to formation of competing structures as the nascent transcript is synthesized. In the latter case it may be that RNA secondary structure is not involved in bringing about the pause and that some other feature of the sequence such as GC content (*6*) is responsible. However in each case a plausible structure can be written if base pairs other than the normal Watson–Crick pairs are al-

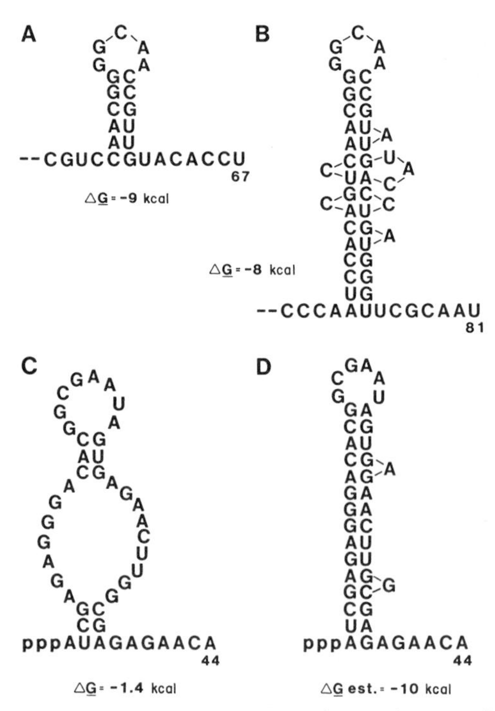

Fig. 6. Predicted RNA structures without pause sites and pause sites without stable Watson–Crick hairpin structures. Structures shown are calculated for full-length T7 A1 transcripts as in Fig. 5: (A and B) predicted stable RNA structures at 67 and 81 nucleotides for the T7ΔD111 transcript; (C) predicted secondary structure upstream of the pause site at 44 nucleotides for the T7ΔD123 transcript; (D) hypothetical RNA secondary structure at the same site shown in C, if A-G base pairs are allowed and are equal in stability to A-U pairs.

lowed. In particular, A-G pairs are known for tRNA and rRNA (*52*) structures, and have also been found in model oligonucleotides (*53*). If such pairs have a comparable stability to the A-U pair, then acceptable secondary structures emerge (Fig. 6D).

While these are plausible explanations, they emphasize the weaknesses inherent in our correlational approach. It remains to be shown that these proposed RNA hairpins actually form in the ternary DNA–RNA polymerase complex. Moreover, to be functionally involved they must presumably form in the time range of 10–20 msec, since the unrestricted elongation rate of RNA polymerase between pause sites is up to 50–100 nucleotides/sec (*17*). There are not techniques now available that allow structural analysis in this time scale of complexes present at nanomolar concentrations. Hence, while it is reasonably certain that some RNA hairpins do lead to pausing of the RNA polymerase during transcription, it is not clear that all pauses are initiated by such hairpins, or what the actual RNA structures are that are involved.

While our mapping studies have only begun to cast light on how DNA and RNA sequences affect transcription elongation, they have been very valuable for other purposes. In particular, the methods we have developed for obtaining RNA polymerase fixed at a single site during elongation provide us with sensitive assays for factors that enhance the microscopic rate of chain elongation or enhance or suppress pausing or termination at particular sites. In addition, we have recently devised a method using limited concentrations of substrates to ''walk'' the polymerase synchronously along the DNA in order to obtain a homogeneous population of polymerase molecules stopped at a particular pause or termination site. This should allow mapping of the RNA transcript structure in actual complexes, as well as footprinting of the contact points of the polymerase along the DNA and RNA chains. Although there are old estimates of the contact areas for elongating polymerase complexes (*54, 55*), these were carried out with heterogeneous mixtures using methods without high resolution and may not have any relevance to the structures involved at pause or terminator sites. Analysis of these complexes should present at least a static picture of the structure at these sites and may provide clues as to how RNA hairpins block the polymerase in its normal elongation sequence.

B. Bacterial Factors Involved in λ Phage Antitermination; Mapping the Components in the *E. coli* Elongation–Termination Complex

Although many of the molecular details are not well understood, it is clear that the folding of the nascent RNA chain impedes the elongation of RNA chains by RNA polymerase *in vitro*. To achieve *in vivo* rates of elongation, formation of RNA structures must be suppressed or some factor(s) must act to block their effect on the elongating polymerase. What are the cellular factors involved? Again, our initial clues derive from studies of termination, in this case the suppression of termination by the N regulatory protein of phage lambda.

In normal λ phage infection, phage transcription is initiated from two major promoters, P_L and P_R, that control leftward and rightward transcription of the λ genome. In the absence of phage gene expression these transcripts are terminated at terminator sites, t_{L1}, t_{R1}, and t_{R2}, respectively. The scheme for rightward transcription is diagrammed in Fig. 7. Expression of the λ phage *N* gene leads to suppression of termination at these sequences and at downstream terminators as well (for review see Refs. *23* and *24*).

This antitermination depends on at least three *E. coli* genes, *nusA, nusB,* and *nusE*. These genes were first identified using mutants that block the action of the λN antitermination function (*56–58*); however, each appears to be an essential *E. coli* gene as well (*25, 26*). The proteins have all been isolated and shown to be active for N antitermination *in vitro* (*59*), although not in a defined system.

The genes for *nusA* and *nusB* have been cloned into overexpression vectors (*60, 61*), which permits isolation of large amounts of each protein for biochemical studies. The *nusE* protein is ribosomal protein S10 (*24*), which is available in reasonably large amounts from purified ribosomes using established methods of isolation (*62*).

The best studied of the *nus* proteins is *nusA*, a 54,000 M_r protein, which was first identified as a factor (L factor) needed for transcription elongation through the *E. coli lacZ* gene (*63*). Genetic and biochemical studies show that *nusA* protein acts in both antitermination and termination processes (*64–66*). Its role as a factor, or cofactor, for termination is best defined for λt_{R2} (*23*) (see Fig. 7), in

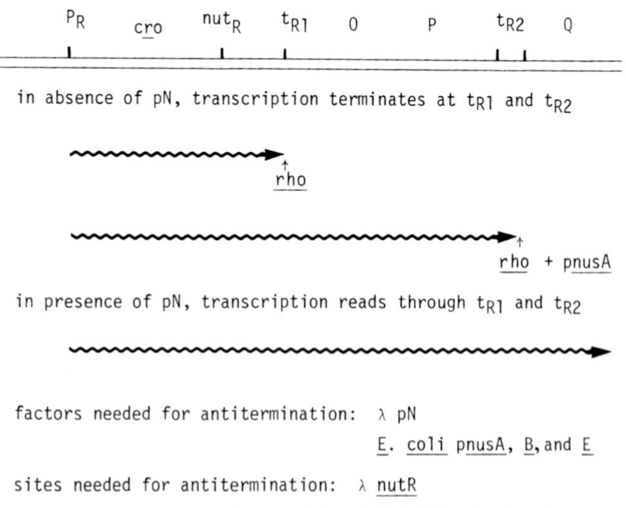

Fig. 7. Diagram of early rightward transcription in lytic infection by phage λ, showing the action of the phage *N* gene antitermination function. See Refs. *23* and *24* for details.

which it has also been shown that purified *nusA* protein enhances rho-dependent termination *in vitro* (*64*). *In vivo, nusA* action is dependent on a specific DNA (or RNA) sequence called box A, which has the average sequence CGCTCTTTA (*67*). Studies of the species specificity of *nusA* action suggest that *nusA* may somehow need to interact directly with the box A sequence in RNA to function in antitermination (*67*).

Second-site suppressor or enhancer mutations for *nusA* mutants have been identified and found to map in the gene for RNA polymerase subunit β, (*rpoB*), in *nusB,* and in the rho gene (*23, 24, 68, 69*). These kinds of mutations can reveal essential protein–protein contacts.

The first of these potential interactions, between *nusA* and RNA polymerase, was confirmed by Greenblatt and his co-workers (*27*), who showed that *nusA* protein binds specifically to core RNA polymerase. From the strength of the binding, and the relatively high level of *nusA* protein in *E. coli,* it is likely that *nusA* protein is normally complexed with the polymerase during RNA chain elongation.

The binding of *nusA* protein to RNA polymerase, taken with genetic evidence for a *nusA–nusB* interaction, suggested that the *nus* proteins might act as a complex bound to RNA polymerase. To detect such interactions we have used protein affinity columns of sepharose coupled either to *nusA* protein or to antibody to *E. coli* core RNA polymerase (*70*). While these studies have not yet detected significant binding of *nusB* protein to either *nusA* protein or core polymerase, they do show significant binding of *nusE* protein to *nusA* protein. In addition, they show, rather unexpectedly, that the rho termination factor also binds well to the *nusA* protein (*70*). Genetic studies (*71*) had suggested that rho might interact with RNA polymerase (*rpoB*), but this interaction could not be reproducibly observed *in vitro* (*70*).

Since *nusA* protein binds to RNA polymerase, binding of *nusE* protein or rho factor to *nusA* might couple these factors to the polymerase. This coupling has been confirmed by using a column of antipolymerase antibody–sepharose (Fig. 8). When core polymerase, rho, and *nusA* proteins are passed through the column, some of the components wash through, but the effluent is then devoid of protein. Elution of the core polymerase with low pH buffer brings off not only core polymerase subunits, but also *nusA* and rho proteins as well.

These results, taken with the genetic studies cited earlier, suggest that the *nus* proteins may act as part of an elongation–termination complex with RNA polymerase during normal RNA chain elongation. We call this hypothetical complex

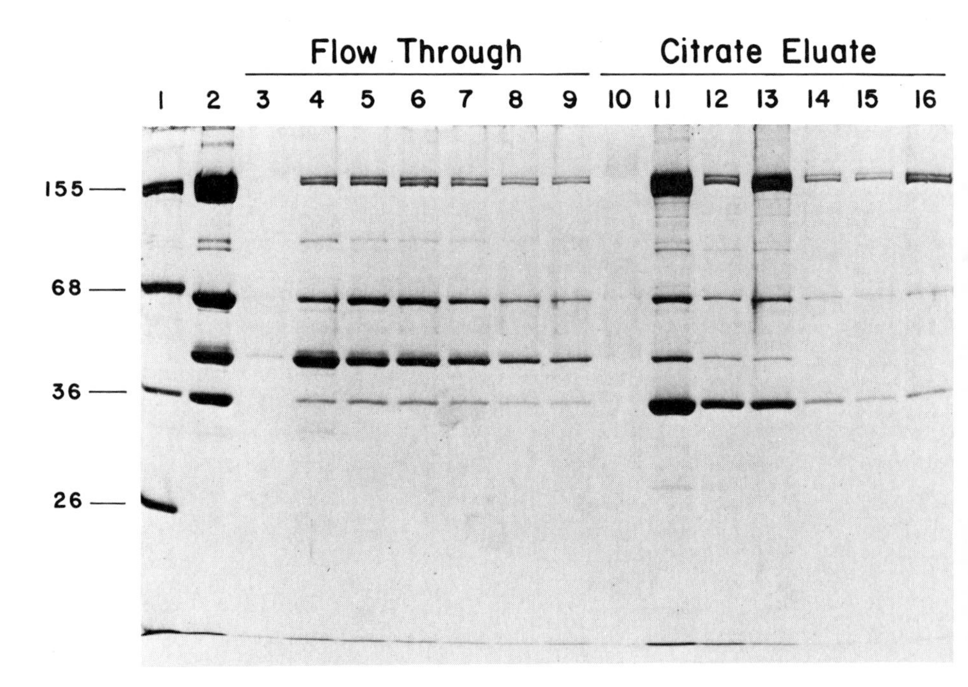

Fig. 8. Coupling of rho protein to *E. coli* RNA polymerase mediated by *nusA* protein. Core RNA polymerase (200 μg) in 1 ml of RB buffer (10 m*M* Tris, pH 8, 5% glycerol, 0.1 m*M* EDTA, 0.1 m*M* dithiothreitol) containing 0.05 *M* NaCl was loaded (1 ml) onto a column of sepharose coupled to purified IgG antibody directed against core RNA polymerase. Experimental details are given in Ref. *70*. After 15 min the column was washed with 10 ml of RB plus 0.2 *M* NaCl, and a mixture of rho protein (25 μg) and *nusA* (25 μg) was loaded in 1 ml of RB containing 0.05 *M* NaCl. After 15 min the column was washed with 10 ml of RB plus 0.05 *M* NaCl and 1-ml fractions, numbered 1 through 10, were collected. Bound proteins were then eluted from the column with 10 ml of 0.1 *M* sodium citrate, pH 3.0, and 1-ml fractions (11 through 20) were collected. Protein components in each fraction were collected by TCA precipitation and analyzed by SDS–polyacrylamide gel electrophoresis (10% gel) followed by staining with Coomassie blue. (Track 1) Molecular weight standards, RNA polymerase β subunits (155,000), α subunit (36,000), bovine serum albumin (68,000), and rabbit triose phosphate isomerase (26,000). (Track 2) Mixture of core RNA polymerase, rho protein (46,000); and *nusA* protein (54,000, however, mobility corresponds to ~65,000 protein, see Ref. *70*). (Tracks 3–9) Proteins from fractions 1, 2, 3, 4, 5, 7, and 9, respectively. (Tracks 10–16) Proteins from fractions 11, 12, 13, 14, 15, 17, and 19, respectively.

a *transcriptosome* in analogy with the multicomponent replication complexes, or replisomes, that carry out replication elongation (*72*). A diagram summarizing possible interactions in the transcriptosome is shown in Fig. 9.

There is no strong reason to believe that rho is part of the transcriptosome

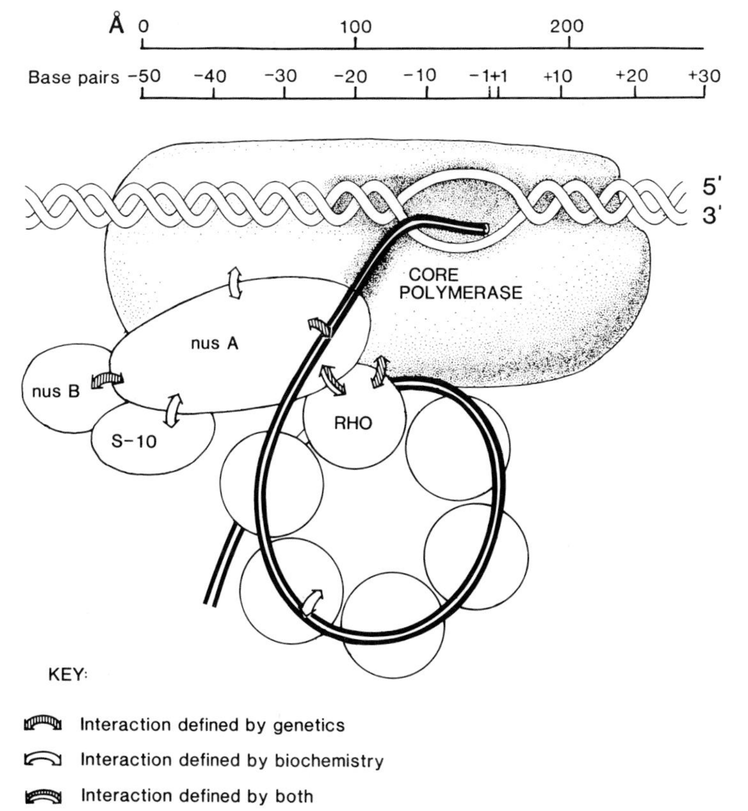

Fig. 9. A model for macromolecular interactions in the *E. coli* transcription elongation–termination complex. Proteins and nucleic acid components are shown approximately to scale. Shaded arrows show possible protein–protein or protein–nucleic acid interactions suggested by genetic studies. Open arrows show interactions detected in biochemical studies. See text for details and references.

during normal elongation. It is usually assumed that rho binds first to RNA sequences and approaches the polymerase only when it is paused at a terminator (*73*). However the coupling of rho to polymerase through *nusA* protein suggests that one must consider the possibility that rho becomes part of the complex even during elongation. It cannot be ruled out *a priori* that rho, like *nusA*, may play a role in both termination and normal elongation.

The model shown in Fig. 9 is inadequate in several respects. As noted before, so far only genetic studies support a *nusB–nusA* interaction. In addition, the model as drawn does not explain why *nusA* function depends on a box A sequence in RNA; *nusA* protein binds to core polymerase in the absence of any RNA (*27*). It may be that encountering the box A sequence changes the structure or binding properties of *nusA* in the transcriptosome, and there is recent preliminary evidence that box A sequence may affect *nusA* protein-dependent antitermination *in vitro* (*66*).

How does formation of a transcriptosome by the *nus* proteins relate to antitermination by λN protein? A plausible model is shown in Fig. 10. It is known that N protein also binds specifically to *nusA* protein (*74*). Hence we assume that N protein may also attach to the transcriptosome to form an N-complex; this interaction probably depends on a DNA (or RNA) sequence called *nutR* (*23, 24*). The resulting N-complex is immune to the action of most terminators.

The fact that rho protein and N protein both bind to *nusA* protein suggests an exciting possibility as to how N binding might block terminator function. In particular, N binding might exclude rho binding or distort that binding to prevent rho from functioning in its termination cofactor role. This possibility, unfortunately, does not explain why N protein can act as an antiterminator for sites that do not require rho factor (*23*). Perhaps these sites also need a release cofactor for efficient action *in vivo* (*75*).

While our understanding of the structural interactions among these compo-

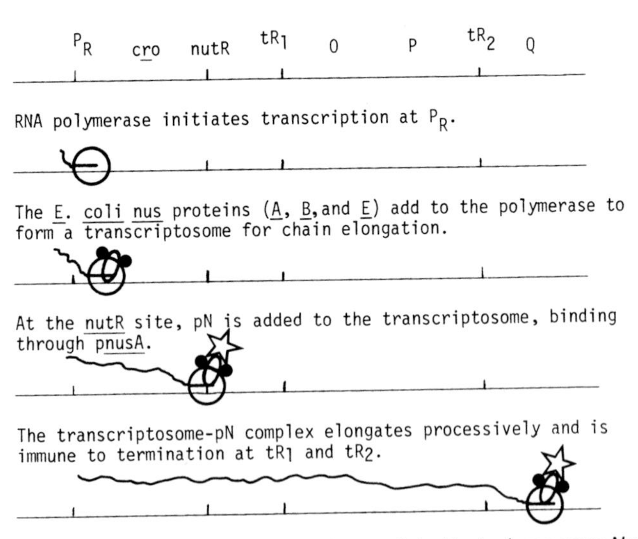

Fig. 10. A plausible model for antitermination mediated by λ phage gene *N* protein. See text for details.

nents of the transcriptional machinery is increasing, our original question remains: What are the factors that suppress pausing and lead to the rapid rates of chain elongation characteristic of cellular transcription? We have carried out extensive biochemical studies on the effects of *nusA* protein on elongation (*60*) and, recently, preliminary studies with purified *nusB* and *nusE* proteins. At saturating concentrations *nusA* protein inhibits transcription elongation from 40 to 90%; two modes of inhibition are implicated. The first is competitive with the nucleoside triphosphate substrates and is not sequence dependent. The second is not competitive with substrates and depends on DNA sequence (*60*). Addition of the *nusE* and *nusB* proteins does not reverse this inhibition under any conditions we have yet studied. Hence it appears that there must be components of the transcription elongation–termination system in addition to the *nus* proteins that are essential in catalyzing normal rates of transciption *in vivo*.

C. A Soluble *E. coli* Cell Extract for the Study of Transcription Elongation and Termination

Although these missing factors could be yet unidentified components of the transcriptosome, more complex possibilities must be considered. For example, if one accepts the general idea that the RNA polymerase is hindered in its elongation by formation of secondary structures in the nascent RNA, then concurrent translation of the RNA might be expected to relieve this hindrance. Hence translation might be the missing element for efficient transcription. It is very clear from electron microscope studies that nascent transcripts are translated with the ribosomes closely following the polymerase. In addition, studies of phenomena such as polarity (*72*) also suggest a functional coupling of transcription and translation, although the details are not well understood.

Because there are so many possibilities, we wanted to obtain a cell-free system involving a total cell extract in which one could study transcription in the presence of concurrent translation. Such an extract should also contain any other components needed for efficient transcription. A very similar approach has been successful in the dissection of the components needed for bacterial DNA replication (*72*). In our initial efforts, all of the "S-30" type extracts that we prepared using conventional methods were unsuitable since they gave high levels of nucleotide incorporation without added DNA, and much of the incorporation was resistant to a mixture of rifampicin and streptolydigin, which should block normal *E. coli* transcription.

Recently, we have obtained a satisfactory cell-free system using a method devised by Kornberg and associates for the study of replication (*76*). Transcription by this system is fully dependent on added template, and synthesis of rather long RNAs can be followed by gel electrophoresis with little nonspecific background. Furthermore, by adding prestarted transcription complexes such as T7 A20 to this system, one can easily measure actual rates of chain elongation by experiments analogous to that of Fig. 3. These kinds of experiments also reveal any effects of the system on pausing or termination.

Our studies using this extract are still in their preliminary stages, but several of the properties of transcription in the system resemble those of the cellular process. In particular,

1. Transcription of coding sequences in T7 DNA is highly dependent on ribosomes and translation. Introduction of an amber mutation into T7 gene *1* reduces transcription about 10-fold compared to the wild-type template.
2. Transcription of T7 leader sequences is stimulated even in regions prior to coding sequences. Elongation rates of 50–100 nucleotides/sec are measured, and there is substantial suppression of pausing at certain sites. This antipause activity has been purified about 10-fold by ammonium sulfate precipitation and DEAE chromatography.
3. Termination at the T7 early terminator, which is rho independent (*77*), is enhanced substantially in such extracts and the site of actual termination is shifted several nucleotides to a site identical to that found *in vivo*. The factor involved is not rho or *nusA* protein or a processing activity but appears to be a new *E. coli* termination factor we have designated tau (*75*). Tau has been purified about 500-fold.

We are optimistic that further study of this soluble system will allow us to identify and isolate the missing components involved in bacterial transcription elongation and termination.

D. A System for the Study of Elongation and Termination by Purified Mammalian RNA Polymerase II

The fact that the elongation and termination phases of bacterial transcription are complex makes it likely that the analogous processes in mammalian cells are at least as intricate. Indeed, it has been known for some time that elongation of

RNA chains by purified mammalian RNA polymerase II is very slow (*29–31*). Even in isolated nuclei, the elongation of transcripts begun *in vivo* is not efficient and normally aborts after a few hundred nucleotides (*32, 34, 35*). Furthermore, little or nothing is known about true transcription termination by RNA polymerase II, and it has even been questioned whether there are true termination sites on the genome for transcripts synthesized by this enzyme.

As some of the basic features of bacterial transcription elongation became clearer, we became interested in looking at the corresponding reactions in mammalian cells. In particular, we wanted to study the elongation and termination of RNA chains through defined DNA sequences using purified mammalian RNA polymerase II. Although this system must certainly be lacking many factors that are involved in transcription *in vivo,* we felt that these studies would be as important for an understanding of mammalian transcription elongation as earlier studies of elongation by purified *E. coli* RNA polymerase, discussed above, had been for the bacterial process. In particular, we expected that these studies would (1) reveal whether transcription elongation by RNA polymerase II is also affected by the secondary structure of the nascent RNA and whether there are unusual features of the reaction; and (2) provide well-defined systems with which one can assay sensitively and specifically for factors that enhance RNA chain elongation and enhance or suppress termination.

Since there are, as yet, no purified systems that give efficient and specific transcription with RNA polymerase II, we chose to bypass the normal initiation reaction entirely. We devised a way of obtaining efficient initiation using highly purified mammalian RNA polymerase at nearly any DNA sequence. This procedure involves the addition of poly(dC) tails to a linear DNA template using calf thymus terminal transferase (*29*). With such "tailed" DNA templates, all of the active RNA polymerase rapidly initiates a chain on the tail just before the beginning of the duplex and then carries out chain elongation through the sequences in question (*29*).

In studying the basic parameters of transcription by RNA polymerase II with these templates, we were surprised to find that the elongation reaction it carries out is highly abnormal. In particular, RNA polymerase II fails to displace its nascent transcript from the template DNA strand and forms, therefore, a stable DNA–RNA hybrid with progressive displacement of the nontranscribed DNA strand (see Fig. 11 for diagram). This is in contrast to what is found with bacterial RNA polymerases, in which the nascent RNA is displaced from the newly formed DNA–RNA hybrid after only 10 or 15 nucleotides. This displace-

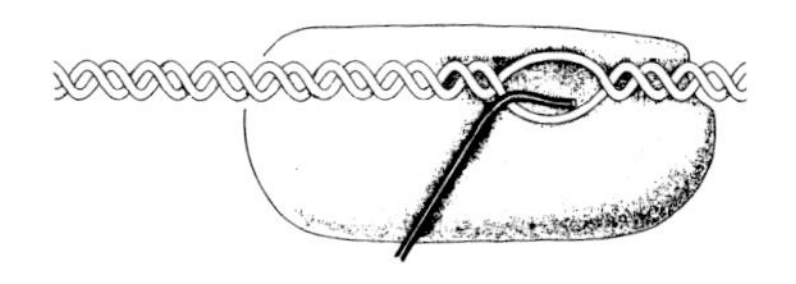

BACTERIAL HOLOENZYME:

Initiation at Specific Promoter Site on DNA
Progressive Displacement of Nascent RNA

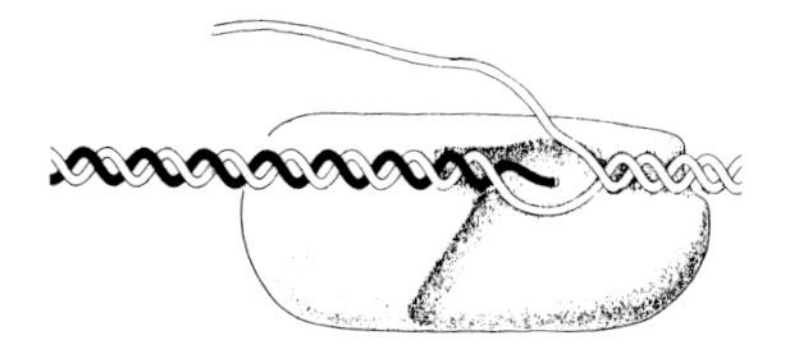

MAMMALIAN POLYMERASE II:

Initiation at Ends or Breaks in DNA
Transcript Remains as DNA–RNA Hybrid

Fig. 11. Comparison of transcription elongation reactions carried out by purified bacterial RNA polymerase and mammalian RNA polymerase II.

ment is known to involve a catalytic function of the polymerase, which can act against a significant free energy gradient (*55*); we will refer to this as a *renaturase* function, since it restores the integrity of the DNA template helix.

This aberrant reaction is a feature of the RNA polymerase protein, not of the template; for example, preparations of calf thymus, human, or *Drosophila* RNA polymerase II give hybrids with all templates tested, while *E. coli* core RNA polymerase and wheat germ RNA polymerase II transcribe tailed templates but make only free RNA (*78*).

We have devised two ways of restoring transcription by RNA polymerase II to the ''normal'' state:

1. When transcription initiates at certain very GC-rich sites, such as an *Sma*I cleavage site, the major fraction of the RNA chains are released normally. This is probably not the usual mechanism for displacement *in vivo* since promoters for RNA polymerase II are not GC-rich at the start sites.
2. We have detected an activity in HeLa cell extracts that seems to catalyze RNA strand displacement plus DNA renaturation during transcription. This activity has been purified over 300-fold, and we call this factor *renaturase* (*79*).

Properties of the partially purified renaturase factor are the following:

1. It allows all of the RNA to be displaced during transcription by mammalian RNA polymerase II with any template we have tested.
2. It does not degrade the RNA, and nascent chains can grow up to at least 5000 nucleotides in length. Electron microscope studies show apparently normal transcription complexes, with full length DNA template molecules.
3. There is no change in the chain elongation rate as measured by growth of chains on RNA gels.
4. Renaturase is not replaced by any of a number of other DNA or RNA binding proteins we have tested so far, including histones, HMG proteins, *Xenopus* topoisomerase I, core proteins from heterogeneous ribonucleoprotein particles, *E. coli* ssb, rho, *recA,* HU proteins, or phage T4 gene 32 protein.

Whether renaturase is a true component of the mammalian transcriptional machinery will require isolation of homogeneous factor and studies of its role *in vivo.*

By using renaturase factor, or by using carefully selected restriction sites, we are now able to follow transcriptional elongation by purified mammalian RNA polymerase II through mammalian or viral genes under conditions such that the RNA transcript is displaced normally. This ability allows us to begin to study the questions posed at the beginning of this section. In our initial studies we have prepared ternary complexes of RNA polymerase II using three nucleotides to run the polymerase into a defined site on tailed template. These complexes are stable and can be used to initiate synchronous transcription through defined sequences.

One such experiment is shown in Fig. 12; transcription has been initiated at a site near the normal initiation site for a human H3 histone gene by cleavage of the DNA and tailing at a *Tha*I restriction site. Because of the GC-rich character of this site, >85% of the transcripts are displaced during transcription (*78*); hence, "normal" transcription elongation is involved. Transcription proceeds through about 500 nucleotides of coding sequence, followed by about 800 nucleotides of downstream sequence before it runs off at a DNA end at around 1300 nucleotides from the start site.

It is evident that, as transcription proceeds through this sequence, there are noticeable transient transcription pause sites, as well as sites at which termination or very long pausing occurs. One such site, at about 480 nucleotides, appears to be a termination release site when release is measured using a nitrocellulose filter assay (*80*).

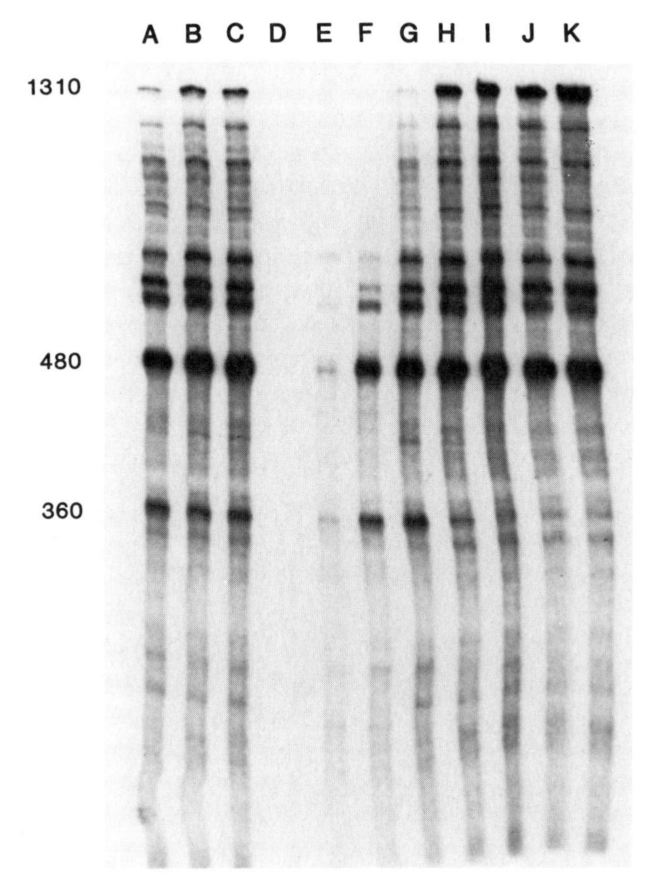

Fig. 12. RNA gel analysis of transcription elongation through a human histone H3 gene by mammalian RNA polymerase II. Plasmid LK189 (6560 bp) contains a 2200-bp DNA insert bearing an H3.3 human histone gene cloned into the *Eco*RI site of pBR322 (*81*). The entire DNA sequence of the H3.3 gene has been determined. This gene is thought to be a pseudogene since there are no obvious promoterlike sequences upstream of the coding region. LK189 was cut with restriction endonuclease *Tha*1 (91 bp upstream from the beginning of the coding sequence and 811 bp downstream from the end of the coding sequence) and poly(dC) tails were then added. The tail downstream from the histone gene was cut off by cleavage with *Acy*1 and *Pst*1 so that transcription in only one direction could be studied. Transcription of this template by purified calf thymus RNA polymerase II gives ~85% free RNA.

Transcription was carried out under ionic conditions described elsewhere (*79*). Calf thymus RNA polymerase II (10 μg/ml) was preincubated at 37°C with template (2 μg/ml), 100 μ*M* GTP, 10 μ*M* UTP, and 10 μ*M* CTP for 10 min to form ternary transcription complexes at

It remains to be shown whether this pausing/termination is affected by RNA secondary structures, as in the case of the bacterial enzyme. However, the extensive termination that is seen seems unlikely to be a normal feature of elongation. Hence it may be that this will provide an assay for factors that suppress this abnormal termination, much like the assay Kung, Spears, and Weissbach (*63*) used for the original isolation of the *nusA* protein from *E. coli*.

We are very encouraged by these results and we think they provide a promising basis for the study of the factors that control transcription elongation and termination by mammalian RNA polymerase II.

References

1. S. Spiegelman, *in* "The Chemical Basis of Heredity" (W. D. McElroy and B. Glass, eds.), p. 232. John Hopkins Press, Baltimore, Maryland, 1957.
2. T. Matsui, J. Segall, P. A. Weil, and R. G. Roeder, *J. Biol. Chem.* **255,** 11992–11996 (1980).
3. M. Samuels, A. Fire, and P. A. Sharp, *J. Biol. Chem.* **257,** 14419–14427 (1982).
4. W. S. Dynan and R. Tjian, *Cell* **32,** 669–680 (1983).
5. M. Rosenberg and D. Court, *Annu. Rev. Genet.* **13,** 319–343 (1979).
6. P. H. von Hippel, D. G. Bear, W. D. Morgan, and J. A. McSwiggen, *Annu. Rev. Biochem.* **53,** 389–446 (1984).
7. R. Kolter and C. Yanofsky, *Annu. Rev. Genet.* **16,** 113–134 (1982).
8. C. Yanofsky, *Nature (London)* **289,** 751–758 (1981).
9. J. P. Ford and M. T. Hsu, *J. Virol.* **28,** 795–801 (1978).
10. N. W. Fraser, J. Nevins, E. Ziff, and J. E. Darnell, *J. Mol. Biol.* **129,** 643–656 (1979).
11. J. R. Nevins, *Annu. Rev. Biochem.* **52,** 441–466 (1983).
12. S. B. Weiss, *Proc. Natl. Acad. Sci. U.S.A.* **46,** 1020–1030 (1960).
13. J. Hurwitz, A. Bresler, and Diringer, *Biochem. Biophys. Res. Commun.* **3,** 15–19 (1960).
14. A. Stevens, *Biochem. Biophys. Res. Commun.* **3,** 92–96 (1960).

---

a defined site 17 nucleotides from the tail–duplex junction. Where noted, complexes were passed through a short Sephadex G-50 column to remove unincorporated nucleotides, and then transcription elongation was carried out in standard ionic conditions in the presence of 800 μM ATP, UTP, GTP, and 100 μM [α-^{32}P]CTP (10^4 cpm/pmol). (Track A) 10-Min elongation in the presence of 800 μM ATP, UTP, GTP, and 100 μM CTP without prior ternary complex formation. (Track B) 10-min elongation of complexes formed but not column purified. (Track C) 10-min elongation of complexes which were column purified. Heparin was added at 4 min to block reinitiation of transcription. (Tracks D–K) Elongation by purified complexes for 0, 1, 2, 5, 15, 30, 45, and 60 min, respectively. Heparin was added at 4 min for samples G–K. Samples were analyzed by electrophoresis on a 4% acrylamide–urea gel followed by autoradiography. Transcript sizes shown were defined by comparison with markers RNAs on the same gel (tracks not shown); the 1310-nucleotide band is the length expected for a full length runoff transcript.

15. M. Chamberlin, *in* "RNA Polymerase" (R. Losick and M. Chamberlin, eds.), pp. 17–68. Cold Spring Harbor Laboratory, New York, 1976.
16. M. Chamberlin, *Annu. Rev. Biochem.* **43,** 721–775 (1974).
17. G. A. Kassavetis and M. J. Chamberlin, *J. Biol. Chem.* **256,** 2777–2786 (1981).
18. H.-F. Kung and H. Weissbach, *Arch. Biochem. Biophys.* **201,** 544–550 (1980).
19. J. Greenblatt, J. Li, S. Adhya, D. Friedman, L. Baron, B. Redfield, H.-F. Kung, and H. Weissbach, *Proc. Natl. Acad, Sci. U.S.A.* **77,** 1991–1994 (1980).
20. R. Kingston and M. J. Chamberlin, *Cell* **27,** 523–531 (1981).
21. H. H. Liebke and J. E. Speyer, *Mol. Gen. Genet.* **189,** 314–320 (1983).
22. K. F. Lech, H. L. Choong, R. R. Isberg, and M. Syvanen, *J. Bact.* **162,** 117–127 (1985).
23. D. Friedman and M. Gottesman, *in* "Lambda II" (R. Hendrix, J. Roberts, F. W. Stahl, and R. A. Weisberg, eds.), pp. 21–52. Cold Spring Harbor Laboratory, New York, 1983.
24. D. F. Ward and M. Gottesman, *Science* **216,** 946–951 (1982).
25. Y. Nakamura and H. Uchida, *Mol. Gen. Genet.* **190,** 196–203 (1983).
26. J. Swindle, J. Ajioka, and C. Georgopoulos, *Mol. Gen. Genet.* **182,** 409–413 (1981).
27. J. Greenblatt and J. Li, *Cell* **24,** 421–428 (1981).
28. R. G. Roeder, *in* "RNA Polymerase" (R. Losick and M. Chamberlin, eds.), pp. 285–329. Cold Spring Harbor Laboratory, New York, 1976.
29. T. R. Kadesch and M. J. Chamberlin, *J. Biol. Chem.* **257,** 5286–5295 (1982).
30. B. Lescure, J. Bennetzen, and A. Sentenac, *J. Biol. Chem.* **256,** 110–11024 (1981).
31. A. Fire, M. Samuels, and P. A. Sharp, *J. Biol. Chem.* **259,** 2509–2516 (1984).
32. R. F. Cox, *Cell* **7,** 455–465 (1976).
33. R. C. C. Huang, M. M. Smith, and A. E. Reeve, *Cold Spring Harbor Symp. Quant. Biol.* **42,** 589–596 (1977).
34. J. Weber, W. Jelinek, and J. E. Darnell, *Cell* **10,** 611–616 (1977).
35. E. Hofer and J. E. Darnell, Jr., *Cell* **23,** 585–593 (1981).
36. N. Maizels, *Proc. Natl. Acad. Sci. U.S.A.* **70,** 3585–3589 (1973).
37. M. Rosenberg, D. Court, H. Shimatake, C. Brady, and D. L. Wulff, *Nature (London)* **272,** 414–423 (1978).
38. L. F. Lau, J. W. Roberts, and R. Wu, *J. Biol. Chem.* **258,** 9391–9397 (1983).
39. W. D. Morgan, D. G. Bear, and P. H. von Hippel, *J. Biol. Chem.* **259,** 8664–8671 (1984).
40. M. Chamberlin, *Fed. Prod., Fed. Am. Soc. Exp. Biol.* **24,** 1446–1457 (1965).
41. F. Martin and I. Tinoco, *Nucleic Acids Res.* **8,** 2295–2300 (1980).
42. V. Brendel and Trifonov, *Nucleic Acids Res.* **10,** 4411–4427 (1984).
43. T. Ryan and M. Chamberlin, *J. Biol. Chem.* **258,** 4690–4693 (1983).
44. N. Neff and M. Chamberlin, *Biochemistry* **19,** 3005–3015 (1980).
45. G. E. Christie, P. J. Farnam, and T. Platt, *Proc. Natl. Acad. Sci. U.S.A.* **78,** 4180–4184 (1981).
46. R. Fisher and C. Yanofsky, *J. Biol. Chem.* **258,** 9208–9212 (1983).
47. F. W. Studier, A. H. Rosenberg, M. N. Simon, and J. J. Dunn, *J. Mol. Biol.* **135,** 917–937 (1979).
48. J. J. Dunn and F. W. Studier, *J. Mol. Biol.* **166,** 477–535 (1983).
49. M. Zuker and P. Stiegler, *Nucleic Acids Res.* **9,** 133–148 (1981).
50. I. Tinoco, P. Borer, B. Dengler, M. Levine, O. Uhlenbeck, D. Crothers, and J. Gralla, *Nature (London), New Biol.* **246,** 40–41 (1973).
51. W. Salser, *Cold Spring Harbor Symp. Quant. Biol.* **42,** 985–1002 (1977).
52. H. F. Noller, *Annu. Rev. Biochem.* **53,** 119–162 (1984).
53. D. J. Patel, S. A. Kozlowski, S. Ikata, and K. Itakura, *Biochemistry* **23,** 3207 (1984).

54. S. Kumar and J. Krakow, *J. Biol. Chem.* **250,** 2878–2884 (1975).
55. J. Richardson, *J. Mol. Biol.* **98,** 565–579 (1975).
56. C. Georgopoulos, *in* "The Bacteriophage Lambda" (A. D. Hershey, ed.), pp. 639–645. Cold Spring Harbor Laboratory, New York, (1971).
57. D. Friedman and L. S. Baron, *Virology* **58,** 141–148 (1974).
58. D. Friedman, M. Baumann, and L. S. Baron, *Virology* **73,** 119–127 (1976).
59. A. Das and K. Wolska, *Cell* **38,** 165–172 (1984).
60. M. Schmidt and M. Chamberlin, *Biochemistry* **23,** 197–203 (1984).
61. J. Swindle and C. Georgopoulos, unpublished studies (1983).
62. C. Kurland, *Methods Enzymol.* **59,** 551–583 (1979).
63. H. Kung, C. Spears, and H. Weissbach, *J. Biol. Chem.* **250,** 1556–1562 (1975).
64. J. Greenblatt, M. McLimont, and S. Hanly, *Nature (London)* **292,** 215–220 (1981).
65. D. F. Ward and M. E. Gottesman, *Nature (London)* **292,** 212–215 (1981).
66. L. Lau and J. W. Roberts, *J. Biol. Chem.* **260,** 574–584 (1985).
67. D. Friedman and E. Olson, *Cell* **34,** 143–149 (1983).
68. M. Baumann and D. Friedman, *Virology* **73,** 128–138 (1983).
69. A. Das, M. Gottesman, J. Wordwell, P. Trisler, and S. Gottesman, *Proc. Natl. Acad. Sci. U.S.A.* **80,** 5530–5534 (1983).
70. M. Schmidt and M. Chamberlin, *J. Biol. Chem.* **259,** 15,000–15,002 (1984).
71. A. Das, C. Merril, and S. Adhya, *Proc. Natl. Acad. Sci. U.S.A.* **75,** 4828–4832 (1978).
72. A. Kornberg, "DNA Replication," Chapter 11. Freeman, San Francisco, 1980.
73. S. Adhya and M. Gottesman, *Annu. Rev. Biochem.* **47,** 967–996 (1978).
74. J. Greenblatt and J. Li, *J. Mol. Biol.* **147,** 16–20 (1981).
75. J. Briat and M. Chamberlin, *Proc. Natl. Acad. Sci. U.S.A.* **81,** 7373–7377 (1984).
76. R. S. Fuller, J. M. Kaguni, and A. Kornberg, *Proc. Natl. Acad. Sci. U.S.A.* **78,** 7370–7374 (1981).
77. M. Kiefer, N. Neff, and M. Chamberlin, *J. Virol.* **22,** 548–552 (1977).
78. R. M. Dedrick and M. Chamberlin, *Biochemistry* **24,** 2245–2253 (1985).
79. C. M. Kane and M. Chamberlin, *Biochemistry* **24,** 2254–2262 (1985).
80. M. Hanna and M. Chamberlin, unpublished studies (1984).
81. D. Wells and L. Kedes, *Proc. Natl. Acad. Sci. U.S.A.* **82,** 2834–2838 (1985).

Lambda (λ) Integrative Recombination: Pathways for Synapsis and Strand Exchange

HOWARD A. NASH

Laboratory of Molecular Biology
National Institute of Mental Health
Bethesda, Maryland

I. Introduction

It has been almost ten years since the first demonstration that lambda (λ) integrative recombination can take place in cell-free extracts (*1*). Since that time the biochemical features of this reaction have been characterized in some detail (*2, 3*). In this chapter, I shall summarize those biochemical findings that I think tell us the most about two critical steps in recombination, the adjoining of two parental DNA helices during synapsis and the breakage–reunion cycle of strand

GENETICS, CELL DIFFERENTIATION, AND CANCER

exchange. Like most biochemists, I believe that basic biochemical mechanisms are used again and again in nature. On this basis, I expect that understanding the λ integrative recombination reaction will yield insight into other site-specific recombinations, such as those that control gene expression and those used by mobile genetic elements such as bacterial transposons and tumor viruses.

To introduce the vocabulary of λ integrative recombination, I begin with a brief review of the components of this reaction. The insertion of viral DNA into the host chromosome is a site-specific recombination, that is, the joining of the phage and host genomes occurs only at specific places (called attachment sites). The recombination crossover takes place within a 15-bp sequence, called the core sequence, that is common to the phage attachment site, *attP*, and the bacterial site, *attB*. Although this homologous region is too short to be recognized by the general recombination systems of *E. coli* and λ, homology is essential for λ integrative recombination. Variants of *attP*, called *saf* mutants, that contains alterations in the core sequence, recombine poorly with *attB* but recombine well with variant *attB*'s that contain the identical *saf* mutation. Thus, λ integrative recombination appears to recognize microhomology. The way in which homology is sensed by the λ system is one of the most fascinating targets of current research. Sequences outside the core are also essential for attachment site function. These sequences, called the arms of the attachment site, are 80–140 bp long in *attP* but only 4–5 bp long in *attB*. As a result of crossing over in the core, the recombinant sites that flank the inserted viral genome are hybrids, containing a long arm from *attP* and a short arm from *attB*.

Only two proteins are directly involved in λ integrative recombination: Int, a viral gene product, and integration host factor (IHF), a host protein. Int binds to two inversely repeated sequences that span the junction between the core and the arms of *attB* and similarly to the core–arm junctions of *attP*. Since Int is thereby positioned near the site of crossing-over one would expect it to play an important role in strand exchange, a hypothesis proved by direct biochemical tests (see below). Int also binds to three regions in the arms of *attP*. This binding must be essential for recombination, since mutations that delete the sites of Int binding in the arms destroy *attP* function. Similarly, IHF binds to three separate regions in the arms of *attP* (*4*); point mutations that eliminate IHF binding also depress recombination (J. Gardner, unpublished observation). One should note that, although recognition of an attachment site by recombination proteins can occur simply (as in *attB*), *attP* function requires a complicated array of bound proteins. This requirement probably reflects the need to form a complex nucleoprotein structure that prepares *attP* to initiate recombination. Such structures have been seen in the electron microscope (*5*) and their consequences for the topology of recombinant products have been delineated (*6*); however, the way in which they activate *attP* remains unknown. For maximum efficiency of recombination, *attP* (but not *attB*) must be present on a supercoiled circle (*7*). It seems attractive to postulate that supercoiling of *attP* is required to promote the formation of a complex nucleoprotein structure, just as supercoiling promotes the formation of protein–DNA structures in chromatin. Indeed, the recent determination of the handedness of nucleosome-like structures at *attP* supports this view (*8*).

II. The Mechanism of Breakage and Reunion

The existence of a homologous core common to *attP* and *attB* suggested to early workers a model for strand exchange in site-specific recombination (*9*) that foreshadowed the methods for *in vitro* recombination that were later adopted by genetic engineers. According to this model, attachment sites would be cleaved within the core by a specific nuclease. Like many restriction nucleases and the nuclease that produces λ termini, the putative recombinase would break the two strands of DNA at different places. The fragmented DNA would then fall apart, leaving a protruding single strand at each end of the break (Figs. 1A and 1B). The complementary protrusions (sticky ends) of *attP* and *attB* would anneal to generate recombinant DNA; recombination would be completed by sealing the joints with a DNA ligase (Fig. 1C). Although this model has many attractive features, the biochemical evidence concerning λ integrative recombination strongly disfavors it. Specifically, biochemical experiments have shown that (1) breakage and reunion do not involve separate nuclease and ligase functions and (2) recombination is not initiated by a double-strand break.

When the phosphodiester backbone of DNA is hydrolyzed by a nuclease, free energy is liberated. Conversely, DNA ligases require a source of chemical free energy (like ATP or NAD) to drive the formation of a phosphodiester bond (*10*). When the proteins that accomplish λ integrative recombination were purified it became clear that no high-energy cofactors were required to accomplish a complete cycle of breakage and reunion. The possibility that the driving force for the reunion step came from the free energy of supercoiling was ruled out by the subsequent demonstration that, under conditions of reduced ionic strength, non-supercoiled DNA could be recombined by purified proteins (*11*). The simplest

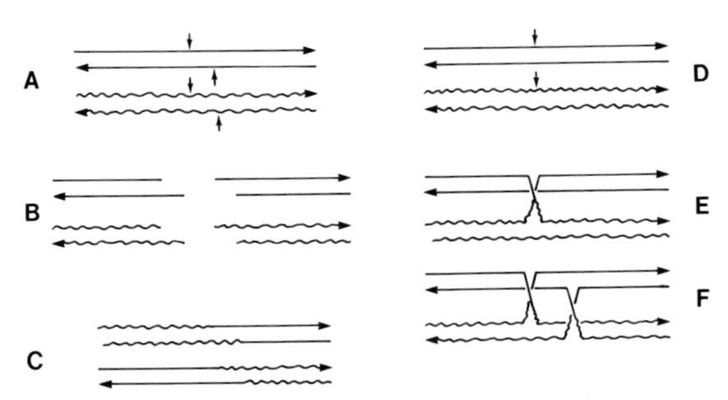

Fig. 1. Models for breakage and reunion. A prototype double-strand break model for recombination is shown at the left. Horizontal arrows indicate strand polarity on the DNA. In (A) the DNA is cut (vertical arrows) with a specific nuclease. The broken ends are shown pulled apart in (B) and rejoined into a recombinant in (C). A prototype single-strand break model for recombination is shown at the right. In (D), recombination is initiated by cutting one strand from each parent. Exchange of these strands produces a Holliday structure (E). A second round of cutting and rejoining on the remaining pair of strands leads to a recombinant (F). In both prototypes, the recombinant is characterized by an overlap region containing one strand from each parent.

conclusion is that breakage and reunion do not use separate nuclease and ligase steps but that the two events are coupled in a way that conserves the free energy of the phosphodiester backbone. This is precisely the mode of action of the class of enzymes called topoisomerases (*12*). These enzymes cleave DNA and form a covalent intermediate between a residue on the enzyme (usually tyrosine) and the phosphate at the site of the break (*13*). Because the phosphate–enzyme bond is of similar energy to the phosphodiester bond, religation of the broken DNA can be achieved by reversal of the cleavage step.

Direct biochemical evidence supports the view that breakage and reunion in λ integrative recombination is accomplished by a topoisomerase. First, Int has demonstrable topoisomerase activity: it can convert supercoiled circular DNA into relaxed closed circles (*14*). Relaxation of supercoils probably reflects an abortive or tangential activity of Int since the activity is very slow, is uncoupled from recombination and does not even require an attachment site. Nevertheless, this activity tells us that a recombination protein carries a breakage and resealing function. Unlike recombination activity, which requires 10–20 monomers of Int per recombinant, the relaxing activity of Int is catalytic, ruling out the possibility that the capacity of Int to reseal DNA is due to a supply of bound cofactor. The second line of evidence concerning topoisomerase action in recombination came from a study of the specificity of Int cleavage. Treatment of an end-labeled fragment of *attP* or *attB* DNA with purified Int results in a small degree of cleavage (*15*). Each strand of the attachment site is cleaved at a unique place within the core and the site of cleavage in one strand is staggered by 7 bp from the site of cleavage in the other strand. Taken together with earlier *in vivo* and *in vitro* results on the position of strand exchange within the core (*16*), this finding establishes both the role of Int in initiating recombination breakage and the exact position of the crossover. That the specific breaks are introduced by a topoisomerase was revealed by the finding that one end of each break in DNA has a free 5′ hydroxy terminus while the other extremity, the 3′ phosphate end, is covalently linked to protein. It should be noted that the chemical polarity of this linkage is not found in other prokaryotic topoisomerases but is identical to that seen in a major class of eukaryotic topoisomerases (*13*). This raises the possibility that topoisomerases might be used in eukaryotes to direct strand exchange as well as to carry out other topological transformations of DNA.

An intrinsic feature of the restriction enzyme–ligase model for recombination described above is the simultaneous cleavage of both strands of each parent to generate cohesive ends. Although this feature can be accommodated in models of recombination that use topoisomerases to carry out strand exchange, the accumulated genetic and biochemical data argue against the initiation of recombination by double-strand breaks. Instead, the data support the mechanism in which the basic event is the cleavage and exchange of one strand from each parent, that is, the formation of a Holliday structure (Figs. 1D and 1E). A distinguishing and attractive feature of the Holliday pathway is that, in contrast to the restriction enzyme pathway, DNA is never completely pulled apart during strand exchange (see Figs. 1B and 1F). The earliest data suggesting the formation of Holliday structures during site-specific recombination came from phage crosses which showed that the position of strand exchange promoted by Int could occasionally be 1 to 2 kb away from the core (*17, 18*). Although the vast majority of crossing-over was confined to the core, a rare class of recombinants behaved as expected if a Holliday structure formed at the core and migrated away from it prior to resolution. The involvement of Holliday structures in integrative recombination has been suggested even more strongly by the high efficiency with which Int protein resolves artifically constructed Holliday junctions into genetically sensible products (*19*). Indeed, the efficiency with which these structures are resolved probably explains why they are rarely observed in genetic experiments and why they fail to accumulate during *in vitro* recombination. The idea that the basic mode of strand exchange involves only one strand from each parent fits nicely with the observation that Int relaxes DNA like a Type I topoisomerase, that is, Int breaks and reseals one strand at a time (*20*). Moreover, specific cleavage of attachment site DNA by Int has been reported to cut only one strand of each duplex (*15*). Recently, however, conditions have been found in which Int can cleave both strands of *attP* (*21*). I believe this phenomenon represents the concerted action of two Int proteins, each cleaving a single strand, and is a biochemical counterpart of the genetic observation that during normal recombination two sets of strand exchanges are usually coupled.

Another line of evidence favoring strand exchange via Holliday structures over cohesive-end joining comes from experiments that focus on the role of homology in recombination. The restriction model of Figs. 1A–1C implies that homology is needed to insure the annealing of the cohesive ends created by double-strand breakage. Such models predict that if homology in the overlap region is disrupted, recombination will abort. As mentioned in Section I, *saf* mutations alter the base sequence in the overlap region. When a *saf* mutation is present in either *attP* or *attB*, recombination *in vivo* is depressed at least 100-fold but when the same *saf* mutation is present in both *attP* and *attB*, recombination is normal. Such mutations therefore demonstrate the need for homology during integrative recombination. However, the behavior of *saf* mutants is not consistent with the predictions of restriction enzyme models for recombination. Specifically, when the DNA from *in vitro* crosses between wild-type and *saf* attachment sites are examined, no broken attachment sites are found (*22, 23*).

One might argue that if the broken attachment sites cannot anneal to form recombinants, they rejoin so as to reconstitute the parental sites. This possibility is refuted by the following experiment (Fig. 2). An artificial *attB* was constructed (Fig. 2C) that contained one strand with a normal overlap region and one strand from a *saf* mutant (*23*). According to the restriction enzyme model for recombination, when this heteroduplex attachment site is cleaved by Int, one fragment will have a cohesive end that can anneal to a wild-type site and one fragment will have a cohesive end that can anneal to a *saf* site (Fig. 2D). Therefore, when this substrate is recombined with wild-type *attP* one expects to get one completed recombinant site and one broken site (Fig. 2E). Alternatively, one could imagine that the failure to reanneal one site might trigger the reversal of the recombination and restore both original parents. However, when such a heteroduplex site was used in a recombination reaction, neither broken attachment sites nor inhibition of recombination was observed (*23*). Instead, the heteroduplex *attB* recombined with normal efficiency with *attP*, producing both *attL* and *attR* in normal yield! This experiment has been performed with two different *saf* mutants, each containing a single mutant base in the overlap region. In each case the mutant

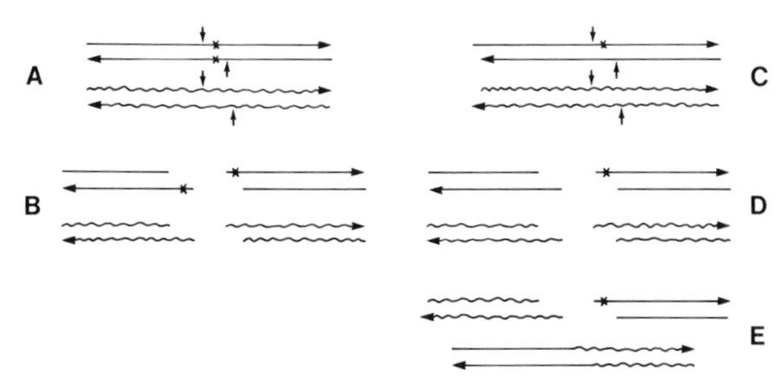

Fig. 2. Predictions of restriction enzyme models for recombination with *saf* mutants. Recombination is depicted as in Fig. 1. On the left is shown a recombination between a wild-type attachment site and a *saf* mutant site (A); the position of the mutation is indicated by an X in each strand. Double-strand breakage of one parent yields ends that are not cohesive with those of the other parent (B). Recombination is predicted either to be stalled at this stage or to revert to the parental configuration (A). Experimental data support the latter alternative (*22, 23*). On the right is shown recombination between a wild-type attachment site and a heteroduplex site that contains one wild-type and one mutant strand (C). Double-strand breakage of the parents yields one set of matching and one set of nonmatching ends (D). The reaction is expected either to proceed, yielding one recombinant and a broken site (E), or to reverse, reforming the parents (C). Experimental data contradict both alternatives (see text).

information was located on either the top or bottom strand of *attB* and recombination was carried out with an *attP* partner that had either a wild-type or a *saf* overlap region. In all cases, when one strand of the heteroduplex *attB* matched one strand of the homoduplex *attP*, recombination appeared normal (*23*). This result strongly disfavors restriction enzyme models for recombination because it refutes a prediction of a basic feature of the model, the need for homology in joining of cohesive ends. In contrast, the result can be accommodated by models that call for formation of a Holliday joint, if only because such models do not make strong predictions about the effect of mismatched bases. The finding that heteroduplex attachment sites that contain three mismatched bases (out of seven) in the overlap region do yield broken attachment sites (*24*) does not mitigate the rejection of double-strand break models. The heavily mismatched sites always yield some completed recombinants (*24*) and in some experimental protocols the proportion of unbroken recombinants can be as high as 90% (J. Gardner, pers. comm.; H. Nash, unpublished observation). This suggests that the breaks in the DNA are the result of secondary cleavage of completed recombinants rather than the failure to join broken parents; it is the least distorted sites that tell us most. In Section IV, I shall present topological experiments that further argue against double-strand breakage models for integrative recombination.

III. A Four-Stranded DNA Model for Synapsis and Strand Exchange

Our understanding of the biochemistry of the breakage and reunion steps in λ integrative recombination opens the way for inquiries into the nature of a more mysterious step: synapsis. During synapsis parental DNAs are juxtaposed so that the broken ends created by Int topoisomerase can be efficiently rejoined to new partners. Because synapsis is transient, we do not yet have a direct biochemical evidence concerning this step; instead we are limited to constructing plausible hypotheses. Three facts about recombination help limit the range of acceptable models: (1) recombination depends on homology within the core; (2) strand exchange is effected by a topoisomerase that breaks one strand at a time; and (3) the ends of DNA created by the breakage step in recombination are not freely diffusible in the interval prior to rejoining. The last fact was deduced from the observation that the products of recombination on supercoiled substrates are themselves supercoiled (*25*); if the broken attachment sites were unconstrained, they would have untwisted prior to rejoining to release the strain of supercoiling. The retention of supercoiling in the recombinant products therefore implies that the recombination machinery holds the broken ends and restricts or guides their movement during strand exchange.

In 1979 (*14*), I proposed a model for synapsis and strand exchange based on these facts (although at the time the requirement for homology was a presumption). Since then, my laboratory has used this model as a working guide for experiments on the topology and biochemistry of λ integrative recombination; Fig. 3 presents an updated version. The model is based on the idea, first put forward by McGavin (*26*), that two homologous stretches of DNA can specifically associate in a four-strand helix (Fig. 3A). From the behavior of stereochemical models of DNA, McGavin concluded that two double helices could wrap around each other's major groove and that homologous base pairs (but not nonhomologous base pairs) would fit nicely within this structure. McGavin's base-pairing scheme is shown in Fig. 3B. I proposed that strand exchange takes place at the edge of this synaptic structure, at the point where four-stranded DNA gives rise to a pair of double helices (*14*). As shown in Fig. 3C and D, Int is imagined to cleave one strand from each parent; the biochemistry of Int cleavage implies that one broken end is a free 5′ hydroxy residue and the other end is covalently attached to Int. The break in each double helix creates a swivel point, and the free end of the duplex (with covalently attached Int) can move with respect to the

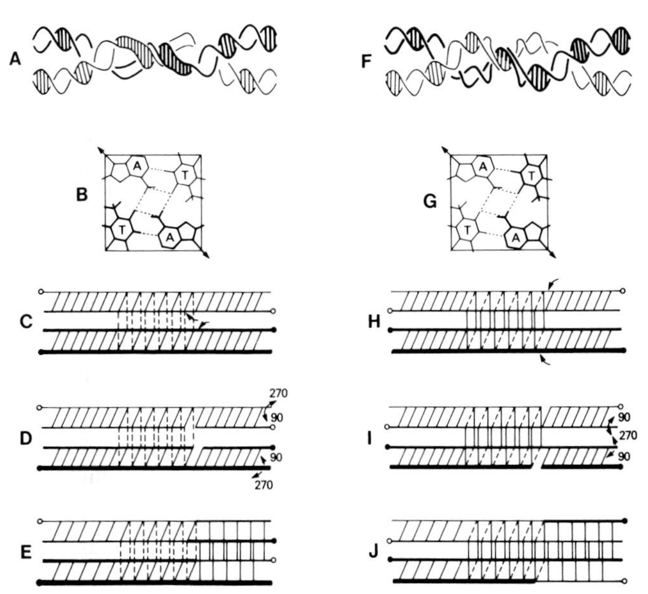

Fig. 3. Strand exchange mechanism with four-stranded DNA synapsis. (A) Two double helices synapsed in a four-stranded helix of the kind proposed by McGavin (*26*). (B) The stereochemistry of the four-stranded regions as seen in a cross section perpendicular to the helix axis. The bases from each of the two parents are marked with heavy and light lines, respectively. The arrowheads indicate the strand whose 5′ end is above the plane of the page. Note the non–Watson–Crick hydrogen bonds that connect the two double helices; these hydrogen bonds (and similar ones formed with G–C pairs) serve to stabilize four-strand helices made from homologous sequences (*26*). (C) The same structure as in (A) but with all the helical character removed for simplicity. The knobs at the end of each strand indicate the 5′ end. Cleavage of DNA by Int at the border of the four-stranded regions, indicated by arrows, initiates recombination. In (D), the rotations required to bring the cut strands adjacent to new partners are indicated. (E) The result of resealing of the newly aligned strands: a Holliday structure is formed. Recombination can be completed by repeating the cycle of cleavage, rotation, and resealing on the remaining pair of strands. (F)–(J) The same mechanism with the minor groove pairing scheme of Wilson (*27*) used to form four-stranded synapsis. Note that in (G) the Watson–Crick hydrogen bonds connect bases from different parents. From Nash and Pollock (*22*).

four-stranded DNA. After rotating only a fraction of a turn the 3′ phosphate end of one parent finds itself opposite the 5′ hydroxy of the other parent and religation occurs (Fig. 3E). Thus, a pair of strands have been exchanged and a Holliday structure created. If the same cycle of breakage, swiveling, and reunion occurs at the other edge of four-stranded DNA on the remaining pair of strands, a reciprocal recombination is completed. Within the framework of this scheme, the observed distribution of recombination break points during λ integrative recombination fixes the size of the four-stranded region at 7 bp. Figures 3F–3J presents an alternative version of this model, in which four-stranded DNA is formed by the pathway first proposed by Wilson (*27*). In this pathway, parental DNAs associate via their minor groove, then melt and reanneal to form two heteroduplexes that wrap around each other as in the McGavin pathway. Int topoisomerase action, swiveling of broken ends and resealing are as before, but the juxtaposition of minor grooves forces a change in the direction of acceptable rotations compared to those allowed with major groove synapsis (cf. Figs. 3D and 3I). This distinction permits topological tests of the two models (see Section IV).

We are in the process of testing the adequacy of four-stranded DNA models for integrative recombination. However, it is not too soon to ask what merit this kind of model might have for other recombining systems. General or homologous recombination is an obvious candidate for the application of this mechanism. Although *in vitro* studies of the conjunction of homologous DNA by *E. coli recA* protein have indicated that the initial act is the pairing of a single strand with a duplex (*28*), the situation is less clear in eukaryotes. In *Saccharomyces,* crossing over is throught to occur by invasion of DNA by an end created by a double-strand break, that is, breakage precedes juxtaposition (*29*). However, in *Ascobolus,* recent genetic experiments have led to the conclusion that a precursor in which two homologous parents are paired precedes not only strand exchange but the formation of heteroduplex DNA (*30*). Four-stranded DNA of the kind described in Fig. 3B fulfills the requirements of such a precursor. Other conservative site-specific recombinations like those promoted by Tn3 resolvase or bacteriophage P1 *cre* protein use topoisomerases to accomplish strand exchange (*31, 32*). However, in these systems the role of homology has not been established and in the case of resolvase, the position of strand exchange would limit the size of four-stranded region to only 2 bp, an unattractive possibility. The next few years should see critical tests of the role of four-stranded DNA in these and other related systems.

Finally, one should consider the applicability of four-stranded DNA models to the transposition of mobile genetic elements. At first glance this might seem far-fetched. First, transposons must associate three pieces of DNA, the two transposon ends and the target, not two pieces of DNA as in the model of Fig. 3. However, as pointed out by Mizuuchi (*33*), strand exchange during the transposition can be formally considered as the sum of two interactions that each involve only one pair of DNAs, one end of the target and one end of the transposon. The second conceptual barrier to relating λ recombination to transposition concerns the role of homology. Transposons almost certainly do not use perfect homology to locate their targets, so the strict version of the model presented in Fig. 3 does not apply. However, one might imagine that transposases have evolved to hold partially homologous or nonhomologous sequences together in a similar arrangement. If this is so, then we can consider the applicability of the model in Fig. 3 to the strand-exchange portion of transposition. Some elements appear to transpose conservatively (*34, 35*), that is, without DNA replication; such "cut and paste" transposons could utilize the topoisomerase cleavage, swivel, and rejoin mechanism of Fig. 3 without alteration. Other mobile elements appear to couple transposition with replication (*36*). For these I propose the following modification. One strand from each end of the transposon would be cleaved by a topoisomerase but the target would be cleaved by a nuclease. Swiveling of the broken helices and rejoining would then create two recombinant joints, each of which would contain a nick and therefore could initiate replication into the adjacent transposon sequences. Thus, the replacement of the topoisomerase cleavage step by a nuclease cleavage in one partner could be the basic distinction between replicative and conservative site-specific recombination.

IV. Topological Features of Integrative Recombination that Distinguish between Models of Synapsis

Closed circular DNAs are topologically invariant. That is to say, because of the continuous nature of the DNA backbone, certain properties of a circular double helix are fixed unless the circle is broken (*13*). The number of times one strand is wound around the other in a circle is one such invariant; this property is called the linking number. The underwound state of negatively supercoiled DNA is reflected in a linking number that is smaller than the linking number of its relaxed closed circular counterpart (*13*). The degree of knotting is another topological invariant. For example, an unknotted closed circle cannot be converted into a knot by simple deformations of the helix such as bending, stretching, or twisting. Each topological state can be altered if and only if one or both strands of the circle are broken and subsequently rejoined. This, of course, is what happens during recombination, and we must expect the act of recombination to leave its footprint on the topological state of DNA. We should therefore be able to tell something about the mechanism of recombination by examining the change in a topological invariant that occurs as a result of recombination. As an example, I have already described how the retention of supercoiling during a cycle of breakage and reunion (*25*) indicated that the recombination machinery holds DNA throughout the period of strand exchange. Recently, T. Pollock and I have refined this experiment so as to measure the precise change in linking number during recombination (*22*). We used a circular substrate in which *attP* and *attB* were oriented as an inverted repeat. Recombination between the sites does not change the size of such a substrate but simply inverts one segment of it with respect to the remainder. By starting with a simple circle of unique linking number, we were able to determine exactly what change took place during recombination. The answer was clear: in those recombinants that remained simple (i.e., unknotted) circles, the linking number changed by two units (I shall deal with the knotted recombinants below). This result is important because it provides quantitative support for models of recombination like those described in the previous section that called for the exchange of strands by two independent cycles of breakage, limited rotation, and rejoining.

Figure 4 diagrams the topological predictions of such recombination mechanisms. In this figure DNA has been reduced to a ribbon in which the edges of the ribbon represent the individual strands of the double helix. Because the two edges of the ribbon are not intertwined, the linking number of this structure is zero. It is important to note that even though all Watson–Crick turns have been omitted from this model, the two surfaces of the ribbon are not equivalent but correspond to the major and minor grooves of DNA. There are three topologically distinguishable ways to juxtapose attachment sites in homologous alignment: major groove to minor groove, minor groove to minor groove, and major groove to major groove. (For the present, we omit discussion of antiparallel synapsis, i.e., juxtaposition of attachment sites such that homologous cores are aligned in opposing directions.) Figure 4 deals only with the last two cases—these correspond to Wilson and McGavin pairing, respectively. In this figure one can see that the rotation of broken strands introduces a twist in the DNA at the point of strand exchange. In addition, the two twisted helices cross each other at the point of strand exchange. To determine the change in linking number produced by these steps one must count the number of strand crossings and sum the effect of each crossing algebraically (*37*). Each crossing is assigned a value of

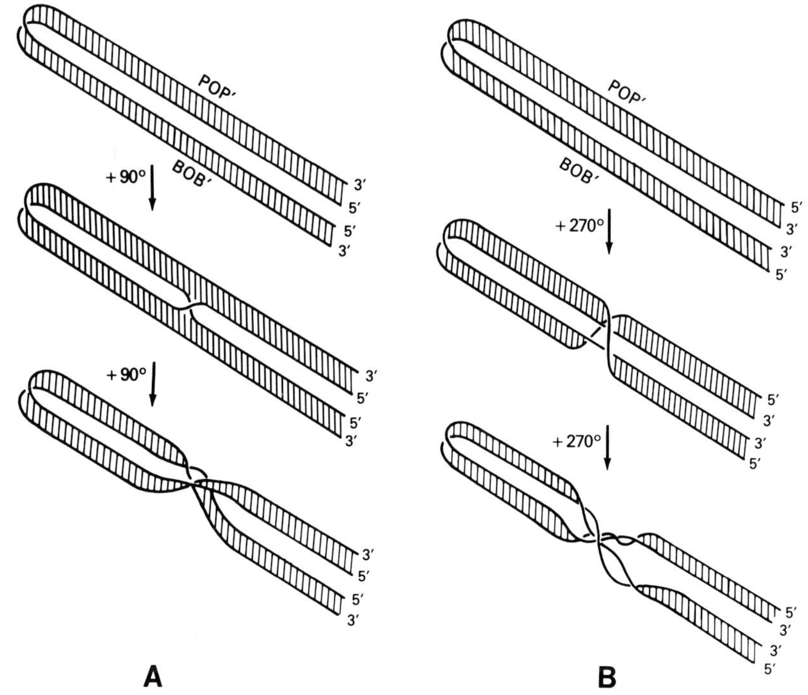

Fig. 4. The topological consequences of recombination. At the top of each panel a segment of a circular substrate containing *attP* (POP′) and *attB* (BOB′) is shown. The core region (O) of each attachment site is juxtaposed across the minor groove surface (A) or major groove surface (B). Note the different orientation of 5′ and 3′ ends of DNA in the two cases (compare with Fig. 3). Breakage, rotation, and resealing of one pair of strands is shown in the central figure of each panel. The bottom figure in each panel shows the completed recombinants that result from a second cycle of breakage, rotation, and resealing.

$\pm 1/2$; the sign of each crossing can be determined from the following convention. First, one specifies an arbitrary direction to the path of the DNA helix. At each strand crossing, an arrowhead is drawn on each strand leaving the intersection.

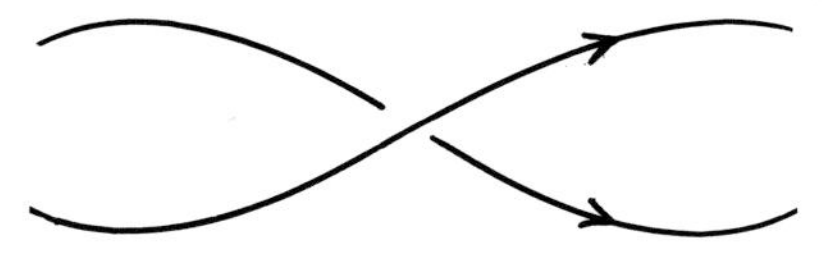

(Note that this convention ignores the chemical polarity of the strands and assigns them the same direction.) The arrowheads can be considered as a pair of vectors arising from the point of strand crossing. One determines the direction of the smallest rotation which will bring the under-crossing vector into alignment with the over-crossing vector. If this rotation is clockwise, the crossing is of positive topological sign; if it is counterclockwise, as in the example above, the crossing has a negative sign. It is important to note that any direction of tracking along the helix axis will yield nodes of the same sign. However, the assignment of a positive value to clockwise rotation of vectors and a negative value to counterclockwise rotation is arbitrary. With the convention used in this paper, the interwindings of the Watson–Crick helix are assigned positive values, and those in the example shown above are assigned negative values.

Using these rules, we can now evaluate the change in linking number caused by recombination. The change introduced by twisting is easily seen to be +1 in the case of +90° rotation and +3 in the case of +270° rotation. The change introduced by the crossing of the twisted helices is harder to follow. Consider first the point of strand exchange following Wilson synapsis and +90° rotation (Fig. 4A). Beside the crossings caused by twisting of the DNA, at the point of strand exchange there are four additional crossings. Two of these involve one strand crossing itself and have no effect on linking number. The remaining two strand crossings are of the same sign and sum to produce in the recombinant a change in linking number of +1. We use the term *node* to describe a place where the helix axis crosses itself; following Wilson synapsis and +90° rotation of the broken strands, the recombinant node has a value of +1. This adds to the change in linking number of +1 introduced by twisting the DNA ribbon to yield a recombinant with a total change in linking number of +2. A similar analysis shows that a recombinant with a linking change of +2 can also be produced following McGavin synapsis (Fig. 4B). Here, the contribution of the recombinant node is −1 and that of the twist introduced by +270° rotation is +3. Thus, both models outlined in Fig. 3 predict topological changes that are in agreement with experiment. In contrast, if one postulates double-strand breakage following either major or minor groove pairing, the simplest motion of DNA leads to a product with a linking number identical to that of the substrate, that is, a change of 0 (*22*). Similarly, pairing the major groove of one parent with the minor groove of the other leads to recombinants with linking numbers changed by +1 or −1 from their substrates (*37*). Thus, in their simplest form, such models are ruled out as acceptable for λ integrative recombination because they are in disagreement with the observed change in linking number. Of course, it is possible to modify models of recombination that call for double-strand breakage or major–minor groove pairing so that they yield an acceptable change of linking number. For example, one could postulate that the recombination machinery adds extra twists to the DNA in the interval between breakage and reunion. Although such motion is not the simplest that one can imagine for these mechanisms, we can not rule out the possibility of its occurrence. Nevertheless, the models described in Fig. 3 are the only class that we have found which generates a linking number change of 2 by moving broken DNA from the proposed synaptic structure in the simplest possible manner. This economy leads us to favor the basic scheme that underlies these models.

How can one distinguish between the two related but acceptable models? In the absence of direct biochemical data, we again turn to topology. Consider the sign of the recombinant node in the two models: minor groove pairing leads to a positive node and major groove pairing leads to a negative node. Unfortunately, in simple circular recombinants there is no way to assess the sign of the recombinant node; it combines with the recombinant twist to yield the overall observed change in linking number. However, if recombination generates a knot, the recombinant node is trapped and its sign can be determined. Figure 5 shows how this can be done. Figure 5A diagrams a nicked circular substrate that contains *attP* and *attB* arranged as a inverted repeat. We imagine that in the process of alignment of these sites, two nodes of the same sign are introduced (Fig. 5B). Then, a recombination node of positive sign can combine with these alignment nodes to generate a trefoil (Fig. 5C). Such a trefoil has three positive nodes. If recombination generates a node of negative sign, the structure in Fig. 5B recombines to generate a simple circle. On a substrate in which alignment has generated the mirror image of Fig. 5B, recombination mechanisms that generate negative nodes can produce trefoils, but these will have three negative nodes. Pollock and Nash studied recombination on such substrates and found that about half of the recombinants were trefoil knots (*6*). Since the substrates were not supercoiled and sites were close together, accidental formation of the structures like those shown in Fig. 5B was unlikely. Instead, Pollock and Nash reasoned that one node was introduced by the obligatory wrapping of an attachment site, presumably *attP*, into a nucleosome-like structure, and the other node simply followed from the need to arrange the sites so that the cores were homologously aligned. Thus, trefoils are a frequent outcome of integrative inversion. Recently, Griffith and Nash have examined the topological structure of these trefoils with the electron microscope (*8*). All the trefoils produced by integrative recombination had three nodes of positive sign. This strongly indicates that recombination

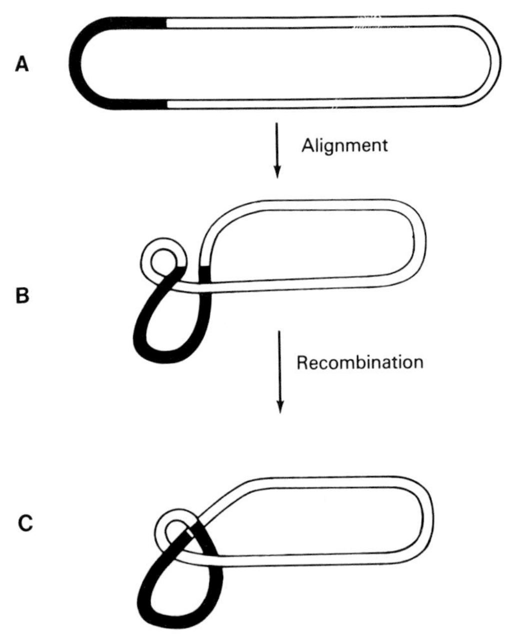

Fig. 5. Site-specific inversion can generate knotted products. (A) An inversion substrate in which *attP* and *attB* are located at the boundaries between shaded and unshaded portions of the circle. In (B) the two sites are juxtaposed so their homologous cores are aligned. The alignment has initiated the entrapment of one segment of the substrate (unshaded) by a loop of another segment of the substrate (shaded). In (C) a strand exchange mechanism that results in a positive recombinant node has completed the entrapment and produced a trefoil.

generates a positive node during strand exchange and therefore rules out models of synapsis that invoke alignment of major groove of one site to major groove of another. Conversely, models of synapsis involving minor groove pairing followed by breakage of one strand from each parent, rotation, and subsequent rejoining agree with all the topological data.

The observation that knotted DNA is a frequent oucome of recombination also comments on the suitability of models of synapsis that invoke alignment of attachment sites in antiparallel fashion. As shown in Fig. 6, parallel alignment easily accounts for the production of both knotted and unknotted recombinants. When parallel alignment produces two nodes of the same sign between the synapsed attachment sites (Fig. 6A), a trefoil can be formed when a third node is added by the strand exchange mechanism. Parallel alignments of homologous cores can also occur with two nodes of opposite sign between the sites; in this

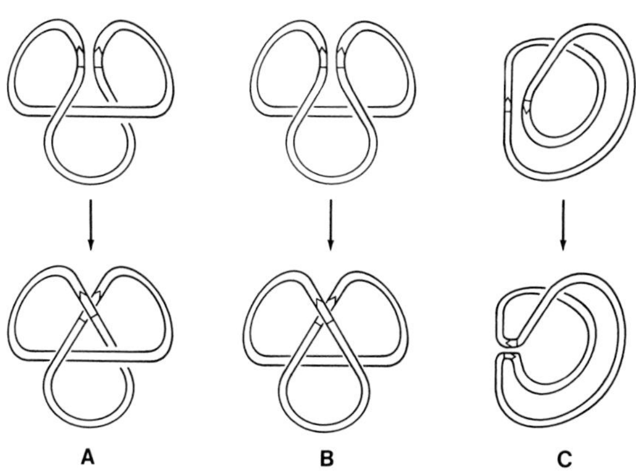

Fig. 6. Parallel and antiparallel synapsis. In each panel an inversion substrate is shown; the attachment sites are marked by arrowheads to indicate the orientation of their core sequences. In (A and B), the sites are aligned in parallel, that is, so that their homologous cores match; in (C) the sites are aligned in an antiparallel fashion. In (A), the alignment has produced two nodes of the same sign; as in Fig. 5, this can lead to the production of a recombinant knot. In (B) the two nodes created by alignment are of opposite sign. This arrangement is topologically indistinguishable from an alignment in which no nodes are formed; recombination can only yield unknotted products. In (C) a single node is required to align the sites; once again, only unknotted products can be formed.

case only simple (unknotted) circular products are generated (Fig. 6B). However, alignment in an antiparallel orientation requires that an odd, rather than an even, number of nodes lie between the sites (Fig. 6C). All simple strand exchange mechanisms on the synaptic structure shown in Fig. 6C generate simple circular products, not trefoils. Unless the mechanism of antiparallel site alignment involves complex motion such that three nodes, rather than one, are trapped, this alignment will not yield knotted products. On this basis antiparallel alignment is disfavored as a suitable model for synapsis during lambda integrative recombination. This is not surprising since the need for homologous recognition between attachment sites has long been interpreted as implying parallel synapsis. However, the topological argument strengthens our mechanistic prejudice.

V. Summary

The biochemical basis of strand exchange during λ site-specific recombination is now understood reasonably well. A topoisomerase function carried by a recombination protein breaks parental DNA at specific places. Like all topoisomerases, the recombination protein covalently joins to the broken DNA and serves as a portable ligase that efficiently carries out rejoining to the broken end of a recombining partner. For λ site-specific recombination the basic event appears to be breakage of one strand from each parent rather than double-strand breakage. This permits exchange to act on one pair of strands at a time, a device that may help to insure that recombination does not abort between breakage and reunion.

The steps that precede strand exchange are less well understood. Recognition of the attachment site by specific binding of recombination proteins is certain but here the problem is to explain the roles of multiple, seemingly redundant protein–DNA interactions. Although the biochemical nature of the early steps in recombination remain to be determined, an analysis of the topological transformations that take place during recombination has been useful in placing limits on what must be occurring. For example, the topological data demand an intimate synapsis that permits very restricted motion of DNA during strand exchange. At least one version of a model for recombination involving four-stranded DNA as the basis for synapsis have been found to concur with all the topological experiments.

Acknowledgment

I thank A. Campbell, P. Kitts, and E. Richet for their comments on the manuscript, and S. Means for preparing it.

References

1. H. A. Nash, *Proc. Natl. Acad. Sci. U.S.A.* **72,** 1072 (1975).
2. H. A. Nash, *Annu. Rev. Genet.* **15,** 143 (1981).
3. R. A. Weisberg and A. Landy, *in* "Lambda II" (R. W. Hendrix, J. W. Roberts, F. W. Stahl, and R. A. Weisberg, eds.), p. 211. Cold Spring Harbor Laboratory, Cold Spring Harbor, New York, 1983.
4. N. L. Craig and H. A. Nash, *Cell* **39,** 707 (1984).
5. M. Better, C. Lu, R. C. Williams, and H. Echols, *Proc. Natl. Acad. Sci. U.S.A.* **79,** 5837 (1982).
6. T. J. Pollock and H. A. Nash, *J. Mol. Biol.* **170,** 1 (1983).
7. K. Mizuuchi and M. Mizuuchi, *Cold Spring Harbor Symp. Quant. Biol.* **43,** 1111 (1979).
8. J. D. Griffith and H. A. Nash, *Proc. Natl. Acad. Sci. U.S.A.* **82,** 3124 (1985).
9. M. Shulman and M. Gottesman, *J. Mol. Biol.* **81,** 461 (1973).
10. I. R. Lehman, *Science* **186,** 790 (1974).

11. T. J. Pollock and H. A. Nash, *in* "Genetic Rearrangement" (K. F. Chater, C. A. Cullis, D. A. Hopwood, A. W. B. Johnston, and H. W. Woolhouse, eds.), p. 41. Croom Helm, London, 1983.
12. J. C. Wang, *J. Mol. Biol.* **55,** 523 (1971).
13. M. Gellert, *Annu. Rev. Biochem.* **50,** 879 (1981).
14. Y. Kikuchi and H. A. Nash, *Proc. Natl. Acad. Sci. U.S.A.* **76,** 3760 (1979).
15. N. L. Craig and H. A. Nash, *Cell* **35,** 795 (1983).
16. K. Mizuuchi, R. Weisberg, L. Enquist, M. Mizuuchi, M. Buraczynska, C. Foeller, P.-L., Hsu, W. Ross, and A. Landy, *Cold Spring Harbor Symp. Quant. Biol.* **45,** 429 (1981).
17. L. W. Enquist, H. Nash, and R. A. Weisberg, *Proc. Natl. Acad. Sci. U.S.A.* **76,** 1363 (1979).
18. H. Echols and L. Green, *Genetics* **93,** 297 (1979).
19. P. L. Hsu and A. Landy, *Nature (London)* **311,** 721 (1984).
20. H. A. Nash, K. Mizuuchi, L. W. Enquist, and R. A. Weisberg, *Cold Spring Harbor Symp. Quant. Biol.* **45,** 417 (1981).
21. P. Kitts, E. Richet, and H. A. Nash, *Cold Spring Harbor Symp. Quant. Biol.* **49,** 735 (1984).
22. H. A. Nash and T. J. Pollock, *J. Mol. Biol.* **170,** 19 (1983).
23. C. E. Bauer, J. F. Gardner, and R. I. Gumport, *J. Mol. Biol.* **181,** 187 (1985).
24. C. E. Bauer, S. D. Hesse, J. F. Gardner, and R. I. Gumport, *Cold Spring Harbor Symp. Quant. Biol.* **49,** 699 (1984).
25. K. Mizuuchi, M. Gellert, R. A. Weisberg, and H. A. Nash, *J. Mol. Biol.* **141,** 485 (1980).
26. S. McGavin, *J. Mol. Biol.* **55,** 293 (1971).
27. J. H. Wilson, *Proc. Natl. Acad. Sci. U.S.A.* **76,** 3641 (1979).
28. P. Howard-Flanders, S. C. West, and A. Stasiak, *Nature (London)* **309,** 215 (1984).
29. J. W. Szostak, T. L. Orr-Weaver, R. J. Rothstein, and F. W. Stahl, *Cell* **33,** 25 (1983).
30. J.-L Rossignol, A. Nicolas, H. Hamza, and T. Langin, *Cold Spring Harbor Symp. Quant. Biol.* **49,** 13 (1984).
31. R. R. Reed and N. D. F. Grindley, *Cell* **25,** 721 (1981).
32. R. Hoess, K. Abremski, and N. Sternberg, *Cold Spring Harbor Symp. Quant. Biol.* **49,** 761 (1984).
33. K. Mizuuchi, *Cell* **39,** 395 (1984).
34. D. E. Berg, *Proc. Natl. Acad. Sci. U.S.A.* **80,** 792 (1983).
35. T. A. Weinert, K. Derbyshire, F. M. Hughson, and N. D. F. Grindley, *Cold Spring Harbor Symp. Quant. Biol.* **49,** 251 (1984).
36. F. Heffron, *in* "Mobile Genetic Elements" (J. A. Shapiro, ed.), p. 223. Academic Press, New York, 1983.
37. N. R. Cozzarelli, M. A. Krasnow, S. P. Gerrard, and J. H. White, *Cold Spring Harbor Symp. Quant. Biol.* **49,** 383 (1984).

Organization and Functional Relationships of Lethal Genes in Mouse *t* Haplotypes

KAREN ARTZT, DOROTHEA BENNETT, AND HEE-SUP SHIN

Memorial Sloan-Kettering Cancer Center
New York, New York

I. Introduction

One approach toward understanding mechanisms of embryonic differentiation and its genetic control in mammals has been the use of mutant genes as tools to dissect early development (*1*). Embryonic lethal genes in the *T*/*t* complex of mouse chromosome 17 have provided an abundant source of such material, since at present about 16 different recessive lethal mutations have been defined as members of that complex (*2*). All of these mutations whose embryological effects have been defined affect very early stages of development, exactly during the time in the first third of gestation when totipotent cells are being channeled into rigidly determined cell lineages. (*3–5*).

We have studied in detail the effects of approximately 10 *t* lethal genes and demonstrated with morphological methods that *t* mutations affect specific cell

GENETICS, CELL DIFFERENTIATION, AND CANCER

types in the early embryo, where their primary effect is the impairment of cell interactions necessary for further development (*6, 7*). Serological and biochemical studies suggested that molecules at the cell surface are altered by *t* mutations (*8, 9*), and we subsequently showed that carbohydrate chains on these molecules are abnormally glycosylated in the presence of *t* mutations (*10–12*). Thus it can be inferred that these mutations either code for mutant cell surface proteins which therefore become abnormally glycosylated and/or that *t* mutations code for aberrant glycosytransferases which add sugars inappropriately to normal cell surface proteins. In any case, the phenotypic effects of each mutation appear to be unique, the expression of each mutation appears to occur during a specific stage of development, and their effects are limited to specific cell types. Thus it has been proposed that *t* lethal mutations may identify wild-type counterparts that switch on and off in some well-regulated sequence to determine cell commitment events during embryogenesis (*13*). Therefore, an important aspect of the *T/t* complex is that it may offer an opportunity to analyze also the organization and regulation of a set of developmentally important genes.

For this reason, in the past several years we have concentrated on genetic and molecular analysis of *t* lethal mutations. These mutations all occur in *t* haplotypes, which are variants of chromosome 17 that occur as polymorphisms in populations of wild mice, where they may reach frequencies of >10% (*14*). All *t* haplotypes contain, usually in addition to one or two embryonic lethal mutations, an assembly of other mutations that affect other processes of differentiation such as tail growth and male germ cell maturation and function (*15, 16*). The factor that affects tail growth, t^{T}, is especially peculiar, interesting, and useful, since it interacts with a common dominant mutation (*T*) in laboratory stocks of mice to produce a unique tailless phenotype in t^{T}/T heterozygotes and thus serves to diagnose the presence of recessive lethal *t* haplotypes, which would otherwise obviously be elusive. The mutations that affect spermatogenesis result in a preferential transmission of the *mutant* chromosome from male heterozygotes at frequencies close to 100% instead of the Mendelian prediction of 50% (*17*). The phenomenon of transmission distortion is no doubt responsible for the maintenance of these lethal chromosomes in wild populations of mice. Although the number of *t* haplotypes defined as different by their content of different complementing lethal *t* mutations is large, as a group they show strong linkage disequilibrium with a much smaller number of major histocompatibility complex (MHC) haplotypes (*18*). This situation has two important aspects. First, since the polymorphism of the MHC is so very extensive in wild-type chromosomes, its

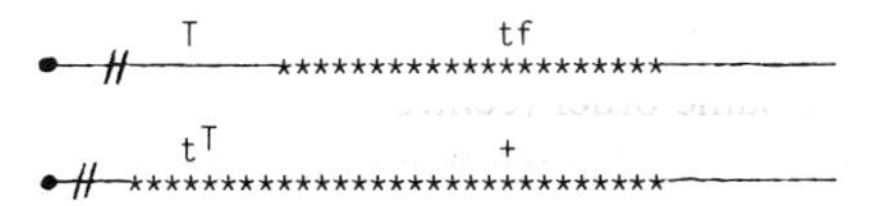

Fig. 1. Crossing over can occur in regions where chromatin peculiar to *t* haplotypes (***) overlaps in heterozygotes. With appropriate markers, such heterozygotes can be used to map genes by classical methods.

restriction in *t* haplotypes suggests that these special chromosomes may be related to one another and descend from a single, or very limited number, of ancestral chromosomes. Second, it reflects the fact that under natural conditions the sets of mutations contained within *t* haplotypes are inherited as a genetic unit, since virtually no meiotic crossing-over occurs between *t* haplotypes and their normal homologs over a distance of about 15 cM that includes the *H-2* complex.

Until quite recently, crossover suppression remained an interesting but enigmatic phenomenon that provided a serious stumbling block to the experimental genetics of the *T/t* complex, since it precluded mapping the lethal and various other mutations in *t* haplotypes by conventional recombinational analysis. The only genetic dissection of *t* haplotypes that was available came from the infrequent crossovers that did take place between *t* haplotypes and wild-type chromosomes. These recombinational events always occur between the markers *qk* and *tf* and result in proximal end recombinant partial *t* haplotypes that retain the t^T mutation responsible for phenotypic interaction with *T*, as well as crossover suppression in the region between t^T and *qk*, and distal end partial recombinants that retain the lethal factor and continue to suppress recombination over the region of *H-2* (*16, 19*). Finally, however, the impediment to genetic analysis was overcome when we showed that crossing over occurs freely between two different complementing *t* haplotypes in the region where "*t* chromatin" is shared (Fig. 1) (*20*). Subsequent experiments that will be discussed below have precisely mapped a number of genes in *t* haplotypes and thus added substantial information to our knowledge of their genetic organization; this information for the first time renders mutations in a gene complex regulating mammalian development susceptible to a molecular analysis of DNA structure.

II. Position of the *H-2* Complex in *t* Haplotypes

Three-point crosses were made to measure recombination between the loci of *T*, *tf*, and *H-2* in heterozygotes for the partial distal *t* haplotype Tt^{s6} and two complete haplotypes, t^{w5} and t^{w32} (see Fig. 2). Our expectation was that these loci would map in the same order (centromere–*tf*–*H-2*) that was well established for wild-type chromosomes, and we therefore scored progeny for recombination between *T* and *tf* and subsequently typed an approximately equal number of recombinants and parental types for *H-2* with serological reagents specific for *H-2* of Tt^{s6} or *H-2* of t^{12}. We found unexpectedly that, although a majority (74 of 119) of recombinants between *T* and *tf* were also recombinant for *H-2*, none of 76 chromosomes carrying parental combinations of *T* and *tf* were also *H-2* recombinants. The obvious interpretation is that in *t* haplotypes the *H-2* complex lies between the loci of *T* and *tf* and thus will maintain its association with parental genes except in the case of a double crossover (*21*).

Thus *t* haplotypes differ from wild-type chromosomes by a major rearrangement that involves an inversion of the chromosome region containing the loci of *H-2* and *tf*. Recombination suppression between *t* and normal chromosomes may be a reflection of this rearrangement, which is common to all *t* haplotypes studied.

III. Orientation of *H-2* in *t* Haplotypes

The rearrangement of *H-2* in *t* haplotypes can be most readily interpreted as a major inversion of at least 10 cM of recombinational distance. To analyze the molecular nature of the rearrangement we identified an *Eco*RI restriction fragment that was polymorphic between two different *t* haplotypes and that mapped to the centromeric end of the *H-2* complex in three independent intra–*H-2* recombinants between different *t* haplotypes (*22;* E. Lader, personal communication). The 22-kb fragment was cloned and shown to be homologous to the *H-2D* region in wild-type chromosomes by genetic mapping in a standard panel of *H-2* recombinant congenic mice. Since the fragment was known already to map at the centromeric end of *H-2* in *t* chromosomes and since the order of *H-2* genes in normal mice is *K–I–D,* this suggested an inversion in the *H-2* complex of *t* chromosomes. To confirm this, we obtained (courtesy of Lee Hood) a unique copy probe to the *Ia* gene, E_α. A *Hin*dIII polymorphism permitted the mapping of this fragment telomeric to the *H-2D* region in *t* haplotypes, thus confirming the gene order as centromere–*H-2D*–E_α in *t* haplotypes, which differs from wild-type chromosomes, in which the order is centromere–E_α–*H-2D* (*23*). Further studies with molecular probes to *TL* and *K* region genes (courtesy of R.

Flavell) have provided data consistent with an inversion of *H-2* in *t* haplotypes, that is, *TL* maps centromeric to *D*, while *K* maps at the distal end of the complex (*24*).

IV. Molecular Structure of the MHC Region of *t* Haplotypes

It was suggested in the introduction to this paper that *t* haplotypes may have a very limited phylogeny, and perhaps even a single origin, since they show such strong linkage disequilibrium with the *H-2* complex. Furthermore, even the different *H-2* complexes that have been defined serologically are now thought to be fairly closely related (*18*). We have attempted to analyze their degree of relatedness at the molecular level by studying their genetic organization with DNA probes homologous to genes coding for class I and class II antigens. The two types of probes have yielded surprisingly different results. With genomic DNA probed with class I sequences on Southern blots it has been shown very clearly that all *t* haplotypes have a remarkably similar structure for these genes and their flanking sequences (*22, 25*). In fact, we have been able to demonstrate that *t* haplotypes, regardless of serological *H-2* type, show a sequence diversity within and around *H-2* class I genes that averages $<5\%$, whereas wild-type chromosomes in inbred mice with different *H-2* haplotypes are divergent by $\sim50\%$. The conclusion from these studies with class I genes was that all *t* haplotypes probably had a common origin in a single ancestral chromosome. Furthermore, since their DNA structure is so similar, it seems likely that the conservation of restriction fragments containing class I sequences must have depended on the suppression of recombination between normal chromosome 17's and *t* haplotypes. This information suggests that as a general rule most sequence divergence may depend primarily on recombinational errors rather than on mutational events. When we examined the same chromosomes with probes to class II *H-2* genes, however, we found quite a different picture. Of 17 different *t* haplotypes analyzed with probes to three I region genes (E_α, E_β, and A_β) and three different restriction enzymes per haplotype we could classify four different types with respect to E_α, seven for E_β, and 11 for A_β (*39*). Thus, especially with respect to A_β, *t* haplotypes are strikingly polymorphic. Furthermore, they also share the strikingly different degree of polymorphism at E_α, which is relatively conserved, and E_β and A_β, which are much more polymorphic, that has already been described for inbred strains (*26, 27*).

The extreme variability of restriction enzyme sites around class II genes is especially interesting, because the very low degree of diversity in restriction sites around class I genes in *t* haplotypes is not the only case in which their relative lack of polymorphism has been demonstrated. For example, in a search for random protein polymorphisms between *t* and wild-type chromosomes eight testicular cell proteins peculiar to *t* haplotypes were identified on two-dimensional gels, none of which, however, varied among different *t* haplotypes (*28*). Similarly, nine random genomic clones isolated from wild-type chromosome 17 DNA were all shown to have *t*-specific restriction fragment polymorphisms, which again were all identical from one *t* haplotype to another (*29*). Finally, protein polymorphisms for known chromosome 17 markers (*glyoxylase 1, TL,* and *Qa*) that lie within the region of crossover suppression in *t* haplotypes have not been detected (*30*). Actually, the only components of the *T*/*t* complex (which is estimated at ~1% of the mouse genome) that do show significant variation, in addition to the *t* lethal mutations themselves, are the genes in the *H-2* region, with class II genes being more variable than class I loci by an order of magnitude. This suggests very strongly that *Ia* genes, and especially A_β, must be subject to some mechanism for generating diversity that is not common to other regions of the chromosome. The nature of this mechanism is of course unknown, but the strong probability is that the general homogeneity of *t* chromosomes is maintained by their restricted opportunities to undergo crossing-over; therefore variability of the *I* region genes is most likely due either to mutation or perhaps to a combination of mutation and selection (*31*).

V. Position of the Lethal Mutations in *t* Haplotypes

Seven different complementing *t* lethal mutations have been mapped relative to one another and to the loci of *T, tf,* and *H-2*. The data show clearly that individual mutations may be separated by very long recombinational distances, as much as 15 cM, although they map primarily in two major clusters: one near the locus of *tf* and another, larger, cluster very closely associated with the *H-2* complex (*32*). In fact we know now that lethal *t* mutations are actually interspersed among the genes of the *MHC*, with two lethal loci mapping between *TL* and *D* (i.e., in the *Qa* region), and another that has not so far been separated from *H-2K*. The positions of two other lethal genes are known to be in or near the

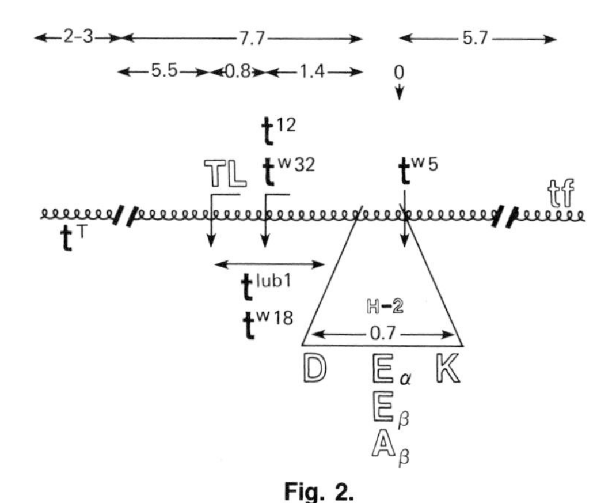

Fig. 2.

TL-Qa region, although their exact positions have not yet been pinpointed (see Fig. 2) (*24*).

These observations have been especially intriguing to us since we have noted that *t* mutations affect only young embryos and spermatogenic cells, whereas *H-2 D* and *K* loci are expressed in all cells except embryos and spermatozoa, and speculated whether the reciprocal temporal expression of these two gene families reflected some genetic or functional relationship (*33*). Furthermore, *I* region genes are also expressed on spermatozoa (*34*), and it has been hypothesized that some of them code for glycosyltransferases (*35*). Thus it is possible that the proximity of *t* and *H-2* genes may provide a functional basis for gene regulation. It is tempting to speculate on the relationship between embryogenesis and *MHC* genes, since the *MHC* contains many class I and class II sequences, for many of which no gene products or functions have been identified. Several recent reports suggest that the *MHC*-linked genes may function during embryogenesis, perhaps in ways similar to "conventional" *H-2* molecules in immune recognition. For example, Rigby and colleagues identified the transcripts of a class I sequence in teratocarcinoma cells (*36*) and showed that a repetitive sequence associated with it is transcribed during embryogenesis (*37*). Furthermore, the *Qa 2.3* antigen has been detected on the surface of preimplantation embryos (C. Warner, personal communication), and *H-2*-linked genes have been shown to affect the rate of early development in mouse embryos (*38*). The ultimate biological significance of all these observations is as yet unclear, but there is no doubt that the molecular analysis of the *T/t–H-2* region of chromosome 17 will help to define genes and gene regulatory mechanisms that function during early mammalian development.

References

1. S. Gluecksohn-Waelsch, *Cold Spring Harbor Symp. Quant. Biol.* **19,** 41 (1954).
2. J. Klein, P. Sipos, and F. Figueroa, *Genet. Res.* **44,** 39 (1984).
3. D. Bennett, *Cell* **6,** 441 (1975).
4. J. Klein and C. Hammerberg, *Immunol. Rev.* **33,** 70 (1977).
5. D. Bennett, *Harvey Lectures Ser. 74,* 1 (1980).
6. D. Bennett, *in* "5th International Conference on Birth Defects" (J. W. Littlefield, J. de Grouchy, and F. J. G. Ebling, eds.), p. 169. Excerpta Medica, Amsterdam-Oxford, 1978.
7. D. Bennett and K. Artzt, *in* "Proceedings of the 6th International Congress of Human Genetics, Part A: The Unfolding Genome" (B. Bonne-Tamir, ed.), Vol. 103, p. 309. Liss, New York, 1982.
8. K. Yanagisawa, D. Bennett, E. A. Boyse, L. C. Dunn, and A. DiMeo, *Immunogenetics (NY)* **1,** 57 (1974).
9. K. Artzt and D. Bennett, *Immunogenetics (N.Y.)* **5,** 97 (1977).
10. C. C. Cheng and D. Bennett, *Cell* **19,** 537 (1980).
11. C. Cheng, K. Sege, A. K. Alton, D. Bennett, and K. Artzt, *J. Immunogenet.* **10,** 465 (1983).
12. B. D. Shur and D. Bennett, *Dev. Biol.* **71,** 243 (1979).
13. D. Bennett, *in* "Current Topics in Developmental Biology" (A. Monroy and A. Moscona, eds.), Vol. 18, p. xii. Academic Press, New York, 1983.
14. D. Bennett, *in* "Origins of Inbred Mice" (H. C. Morse, ed.), p. 615. Academic Press, New York, 1978.
15. D. Bennett, *Science* **144,** 263 (1964).
16. M. F. Lyon and R. Meredith, *Heredity* **19,** 301, 313, 327 (1964).
17. D. Bennett, A. K. Alton, and K. Artzt, *Genet. Res.* **41,** 29 (1983).
18. D. Nizetic, G. Figueroa, and J. Klein, *Immunogenetics (N.Y.)* **19,** 311 (1984).
19. D. Bennett, K. Artzt, J. Cookingham, and C. Calo, *Genet. Res.* **33,** 269 (1979).
20. L. M. Silver and K. Artzt, *Nature (London)* **290,** 68 (1981).
21. K. Artzt, H-S. Shin, and D. Bennett, *Cell* **28,** 471 (1982).
22. H-S. Shin, J. Stavnezer, K. Artzt, and D. Bennett, *Cell* **29,** 969 (1982).
23. H-S. Shin, L. Flaherty, K. Artzt, D. Bennett, and J. Ravetch, *Nature (London)* **360,** 380 (1983).
24. H-S. Shin, D. Bennett, and K. Artzt, *Cell* **39,** 573 (1984).
25. L. M. Silver, *Cell* **29,** 961 (1982).
26. M. Steinmetz, K. Minard, S. Horvath, J. McNicholas, J. Frelinger, C. Wake, E. Long, B. Mach, and L. Hood, *Nature (London)* **300,** 35 (1982).
27. R. R. Robinson, R. N. Germain, D. J. McKean, M. Mescher, and J. G. Seidman, *J. Immunol.* **131,** 2025 (1983).
28. L. M. Silver, J. Uman, J. Danska, and J. I. Garrels, *Cell* **35,** 35 (1983).
29. D. Rohme, H. Fox, B. Harrmann, A-M. Frischauf, J-E. Edstrom, P. Mains, L. M. Silver, and H. Lehrach, *Cell* **36,** 783 (1984).
30. K. Artzt, P. McCormick, and D. Bennett, *Cell* **28,** 463 (1982).

31. K. R. McIntyre and J. G. Seidman, *Nature (London)* **308,** 551 (1984).
32. K. Artzt, *Cell* **39,** 565 (1984).
33. K. Artzt and D. Bennett, *Nature (London)* **256,** 545 (1975).
34. G. J. Hammerling, G. Mauve, E. Goldberg, and H. O. McDevitt, *Immunogenetics (N.Y.)* **1,** 428 (1975).
35. C. R. Parish, C. H. O'Neill, and T. J. Higgins, *Immunol. Today* **2,** 98 (1981).
36. P. M. Brickell, D. S. Latchman, D. Murphy, K. Willison, and P. W. J. Rigby, *Nature (London)* **306,** 756 (1983).
37. D. Murphy, P. M. Brickell, D. S. Latchman, K. Willison, and P. W. J. Rigby, *Cell* **35,** 856 (1983).
38. S. B. Goldbard, K. M. Verbanac, and C. M. Warner, *Biol. Reprod.* **26,** 591 (1982).
39. K. Artzt, H.-S. Shin, D. Bennett, and A. Dimeo-Talento, *J. Exp. Med.* (in press) (1985).

Control of Gene Expression during Terminal Cell Differentiation*

R. A. RIFKIND, M. SHEFFERY, AND P. A. MARKS
DeWitt Wallace Research Laboratory
Sloan-Kettering Institute
Memorial Sloan-Kettering Cancer Center
New York, New York

I. Introduction

Terminal cell differentiation in the erythropoietic lineage constitutes one of the best characterized developmental systems explored in higher organisms. The

*Research summarized in this review from the authors' laboratory was supported, in part, by the National Cancer Institute (PO 31768 and CA 08748), and, in part, by the Bristol-Myers Cancer Grant Program. M.S. was supported, in part, as a Miriam and Benedict Wolf Fellow in Molecular Genetics.

morphological and biochemical changes, as well as molecular events including the synthesis of differentiation-specific mRNAs (globin mRNAs) and other differentiation-specific products, have been described (*1–4*), and molecular probes for the definition of specific gene expression are available.

For experimental studies, murine erythroleukemia cells (MELC), a virus-transformed precursor which, in terms of its developmental status, appears to be at the colony-forming unit for erythropoiesis (CFUe) stage of erythropoietic development, provide significant advantages in the study of regulatory mechanisms implicated in differentiation (*1*). These cloned, transformed cells, arrested in their normal differentiation pathway, provide an opportunity to define the cellular and molecular phenotype of this developmental stage.

Upon exposure to any of a number of chemical agents, MELC are induced to initiate their normal developmental program, providing an additional opportunity to study at the cellular and molecular level those events which accompany and may control the initiation of terminal cell differentiation in this lineage.

MELC display yet other advantages as an experimental model including the availability of a variety of inducing agents capable of initiating different patterns of expression of the features of induced differentiation (*1, 5–8*), as well as a number of variant cell lines that express sensitivity or resistance to inducing agents or to agents capable of inhibiting the process of induced differentiation (*9*).

The program of induced differentiation initiated by chemical inducing agents is complex, including, for example, morphogenetic changes (*10*), changes in cell membrane properties (*11–13*), cyclic nucleotide metabolism (*14*), iron transport, and heme synthesis (*4*). In this review we will concentrate on two aspects of induced MELC differentiation that have been examined in our laboratories: (1) the control of globin gene expression at the transcriptional level during induced differentiation; and (2) mechanisms regulating the initiation of terminal cell division, which is a characteristic feature of both normal and induced erythroid cell differentiation.

II. DNA Structure, Chromatin Configuration, and Gene Expression during Induced Differentiation

The globin genes provide a model system for the study of mechanisms controlling the cordinated expression of unlinked, differentiation-specific genes during differentiation. During induced differentiation there is a 10- to 20-fold increase in the rate of accumulation of globin mRNAs, regulated, to a significant degree, by increased transcription (*15, 16*), although posttranscriptional control has also been implicated (*17*). Two hemoglobins, Hb^{maj} (containing α- and β^{maj}-globin polypeptides) and Hb^{min} (α- and β^{min}-polypeptides) are synthesized in differentiating MELC. There is a low, constitutive, level of Hb^{min} synthesis in uninduced MELC, suggesting a low level of expression of the α- and β^{min}-globin genes in these precursor cells (*8*). Different classes of inducing agents have differential effects on the several globin genes; polar planar inducers, such as Me_2SO and hexamethylene bisacetamide (HMBA), induce more Hb^{maj} than Hb^{min} while fatty acids (butyric and other short-chain fatty acids) initiate accumulation of roughly equal amounts of both proteins. Hemin induces, predominantly, the accumulation of Hb^{min} (*8, 18*).

Physical maps of the murine α- and β-globin gene domains are available (*19, 20*), and both cDNA and genomic clones have been prepared. Seven loci have been identified within a 60-kb β-globin gene domain (chromosome 7), corresponding to expressed β-globin genes or pseudogenes (*21*), while a 40-kb region contains the α-embryonic globin genes and the α_1- and α_2-adult globin genes (chromosome 11) (*22*).

A. Control of Globin Gene Expression during Induced Differentiation

There is considerable evidence, from a number of cell systems, that globin gene expression during development is regulated at the level of gene transcription (*23–28*). In MELC, a 10- to 20-fold increase in β^{maj}-globin gene transcription has been demonstrated during induced differentiation, employing the nuclear RNA chain elongation assay (*15*); transcription initiates at the cap site, occurs predominantly off the coding strand, and extends about 1.5 kb beyond the poly(A) addition site before termination. Transcription of the α_1-globin gene also increases 10- to 20-fold during induction, initiating at or near the capsite and terminating in a region 50–250 bp 3′ to the polyadenylation site (*29*). Following exposure to the inducer HMBA, α_1-globin gene transcription is detected within two to three cell cycles (about 36 hr) while β^{maj}-globin gene transcription is detected later, about 12 hr after initiation of α-globin gene transcription.

B. DNA Structure and the Control of Globin Gene Expression

The role of DNA methylation has been examined as a potential mechanism for the regulation of globin gene expression during induced differentiation. We have examined the pattern of cytosine methylation in the nucleotide sequence CCGG, using the methyl-sensitive isoschizomer-pair of restriction enzymes *Msp*I and *Hpa*II, and other restriction enzymes (*16*). Relatively few such potentially methylated sites in the MELC globin gene domains can be found; of the sites near the β^{maj}-globin gene, one is fully methylated, one partially methylated, and one unmethylated in uninduced cells. Most, but not all, sites near the α-globin genes are unmethylated in uninduced cells. No change in the pattern of methylation about either gene is observed during HMBA-mediated induced differentiation. Both α_1 and β^{maj}-globin genes, however, display distinctly more methylation in nonerythroid mouse tissues than they do in MELC. Thus, it would appear that a pattern of globin gene–related methylation appropriate for expression of these genes has been established in MELC at an earlier stage in their developmental history; no further change in globin gene–related methylation, within the limits of this assay, appears to be required for the initiation of transcription.

C. Chromatin Structure and the Regulation of Globin Gene Expression

Based upon evidence from a number of systems it appears likely that changes in chromatin structure, dependent upon chromosomal proteins, play a significant role in regulating the transcription of developmentally specific genes (*30, 31*). Several features distinguish the chromatin of active chromosomal regions from bulk chromatin and appear to reflect aspects of nuclear protein configuration important for transcriptional activity or quiescence. We have examined several of these features with respect to the expression of globin genes during induced MELC differentiation, including the sensitivity of chromatin to digestion by DNase I, the nucleosome configuration of sequences within and surrounding the globin gene domain, and the appearance of discrete sites that are hypersensitive to digestion by a number of endonucleases including DNase I and S_1 nuclease. Each of these aspects of chromatin configuration has been examined in detail during HMBA-mediated induced MELC differentiation.

1. DNase I Sensitivity

Relative accessibility to DNase I digestion appears to be characteristic of the chromatin of active or potentially active transcriptional domains (*31*). We have demonstrated (*16*) that the β^{maj}- and α_1-globin gene–associated chromatin regions are more sensitive to digestion by DNase I than is the *Ig*α (immunoglobulin) gene (which is not expressed by cells in the erythroid lineage), in both uninduced and induced MELC. As already suggested in the case of the globin gene methylation pattern, it appears that the globin gene–associated chromatin domains have been selectively modified during the developmental history of the erythroid lineage prior to the MELC stage of differentiation, placing the globin-related chromatin in a potentially active, although relatively unexpressed, configuration. This feature, like the methylation pattern, is stably propagated in uninduced MELC (*32*) and constitutes another aspect of the molecular phenotype of erythroid cells that have differentiated to the CFUe-like (MELC) stage of erythroid cell development.

2. Nucleosome Configuration

Using a chromatin fraction procedure based upon micrococcal nuclease digestion and solubility in EDTA-containing solutions, we have demonstrated, on the one hand, that genes that are not expressed in the erythroid lineage, such as the immunoglobulin and albumin genes, distribute preferentially into the soluble rather than insoluble fractions and are organized into canonical nucleosomes (*33*). On the other hand, both α_1- and β^{maj}-globin genes are enriched in the insoluble fraction and are organized into structures partially devoid of nucleosome configuration in uninduced MELC at a time when these genes are transcriptionally inactive. The domain of nucleosome disruption begins approximately 300 bp 5′ to the cap site of the β^{maj}-globin gene and 1500 bp 5′ to the α-globin gene. (Whether these structural differences between the α and β globin genes bear a relationship to known functional differences between these genes remains to be determined.) More distant flanking sequences partition into the soluble fractions and are organized in classical nucleosome structures. Thus, it would appear, a state of partial nucleosomal disruption in the globin gene domains is yet another aspect of the molecular phenotype of cells at the MELC (CFUe-like) stage of erythroid cell differentiation.

Following HMBA-mediated induction of MELC differentiation and the onset of globin gene transcription, there is a modest increase in the level of nucleosome disruption (most pronounced at the 5′ ends of the globin gene domains) but the basic structural features are unchanged. In contradistinction to the globin genes, ribosomal genes, which are inactivated during induced differentiation (*15, 34*), are progressively reconfiguraed into a nucleosome chromatin pattern (*33*).

3. Nuclease Hypersensitivity Sites

Not all features of chromatin structure associated with transcriptional activity are found in the uninduced MELC. We have obtained evidence for alterations in chromatin structure that are specifically associated with inducer-mediated activation of globin gene transcription (*16, 35*). Sites displaying a six- to 10-fold increase in DNase I sensitivity (hypersensitivity sites) appear in chromatin regions near the 5′ end of the α_1 and β^{maj} genes during HMBA-induced differentiation. Present evidence suggests, however, that changes in chromatin structure that are revealed by nuclease probes during induced differentiation are more complex than simply the reconfiguration of regions at the 5′ end of the globin genes. There is a DNase I hypersensitive site, located within the second intron (IVS-2) of the β^{maj} gene, which can be detected in uninduced MELC; this hypersensitivity site disappears during HMBA-mediated differentiation, to be replaced by the new hypersensitivity site, which lies 5′ to the β^{maj} cap site. This complex change in chromatin configuration takes place normally in a coordinated fashion, just before initiation of globin gene transcription. The two changes in chromatin configuration can, however, be dissociated. In the R1 MELC variant, which is resistant to the inducing effects of HMBA on globin gene expression and commitment (*36*), HMBA induces the disappearance of the hypersensitivity site located in IVS-2 but fails to generate the new 5′ site and likewise fails to initiate transcription at the β^{maj} gene (*35*).

There is also a complex pattern of chromatin reconfiguration that occurs at the α_1-globin gene domain during induced differentiation (*29*). Uninduced MELC display overlapping DNase I and S_1 nuclease–sensitive sites located 5′ to the α-globin gene cap site. During induced differentiation the nuclease sensitivity of these sites increases and new, nonoverlapping sites develop, one located about 300 bp 5′ to the α_1-globin cap site and the other virtually coincident with the cap site itself. These changes in nuclease sensitivity do not occur in the HMBA-resistant variant, suggesting that there exist significant differences between the chromatin-associated events that take place in the α- and β-globin gene domains during induced differentiation.

Taken together, these observations lead us to propose the model illustrated in

Fig. 1 for the changes in chromatin structure associated with erythroid cell differentiation and induced globin gene expression in MELC. In undifferentiated precursors globin genes are, presumably, organized into nucleosome arrays, which are transcriptionally inactive. During those developmental stages of the erythroid lineage that precede the CFUe-like (MELC) stage of differentiation,

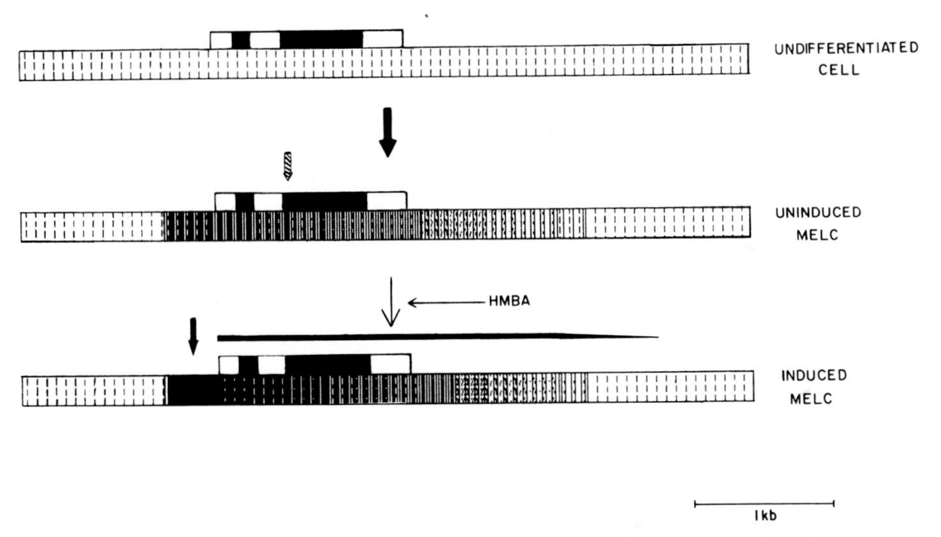

Fig. 1. Changes in nucleosome and chromatin in the region of the α^{maj}-globin gene during erythroid cell differentiation. The configuration of chromatin around the β^{maj}-globin gene in undifferentiated cells (putative stem cells) in uninduced MELC and in MELC induced to terminal erythroid differentiation are shown. During erythropoiesis the β^{maj}-globin gene chromatin undergoes a series of structural alterations that appear preparatory for transcription. The β^{maj}-globin gene is indicated by the box containing the two introns (black) and three exons (white) of the globin genes. The structure of the chromatin region across the β^{maj}-globin gene is indicated by a strip below the gene, which is incrementally shaded in direct proportion to the relative loss of canonical nucleosome structure. The structure of the globin gene–associated chromatin of the undifferentiated cell is hypothetical. Note the polar nature of the changes in chromatin structure: regions at the 5′ end of the gene are twice as likely as regions at the 3′ end of the gene to assume the nonnucleosomal configuration. The small hatched vertical arrow indicates the DNase I hypersensitive site, located at the 5′ end of the second intron, in uninduced MELC. Induction of MELC with HMBA results in further chromatin structure alterations along the transcribed domain leading to increased nucleosome disruption. The small solid vertical arrow indicates a DNase I hypersensitive site that appears at the 5′ end of the gene, 200 bp upstream of the cap site, after induction. The tapered horizontal line drawn above the induced gene represents the approximate dimensions of the transcribed domain. This figure is modified from Ref. *33*.

the globin gene–assocaited chromatin domains undergo a series of structural alterations as a prelude to transcriptional activation. These stably propagated features are recognized by an increased sensitivity to DNase I (*15, 16, 32*) and micrococcal nuclease (*37, 38*), partial disruption of nucleosome structure concentrated at the 5′ end of the gene (*33*), and a unique pattern of nuclease hypersensitivity sites (*16, 29, 35*). These features, as well as the distinctive hypomethylation pattern, constitute the present state of our understanding of the molecular phenotype of the MELC (CFUe) stage of erythroid cell differentiation. Induction of terminal differentiation in these cells, by HMBA or other inducing agents, results in further chromatin structural changes near the globin genes, reflected in establishment of novel DNase I and S_1 nuclease hypersensitivity sites (*15, 16*) and a modest increase in the level of nucleosome disruption (*33*) associated with the onset of globin gene transcription.

III. Commitment to Terminal Cell Division

MELC can be maintained essentially indefinitely in suspension culture under appropriate conditions (*1*), displaying the unlimited proliferative capacity of this transformed erythroid precursor cell. A variety of inducing agents, including HMBA, initiate not only the transcription of the differentiation-specific globin genes, but also the complex process of terminal cell division which is characteristic of induced differentiation of this transformed precursor and also of the normal terminal stages of differentiation in the erythroid lineage (*1*). An assay for examining the commitment of MELC, at the single-cell level, to terminal cell division and the loss of capacity for cell proliferation has been developed (*39, 40*). MELC, first exposed to inducer in suspension culture, are then grown as a single cell suspension in semisolid medium in the absence of inducer and the growth of each cell scored by examining the progeny colonies after several days. Small colonies of <32 cells, containing hemoglobinized cells, are the progeny of those MELC that were committed to the program of terminal cell division (and erythropoietic differentiation) during the period of their exposure to inducer. The rate and level of commitment is dependent upon the nature, as well as on the duration of exposure to and concentration of the inducing agent (*41*). Commitment can be detected as early as 12 hr in culture with HMBA and almost the entire population can become committed to terminal cell division by 48 hr in culture with this agent (*41*).

A. Commitment to Terminal Cell Division: a Multistep Process

That commitment is a multistep process is suggested by a number of lines of evidence. Different inducers initiate commitment with different kinetics (*8, 41, 42*), while variants of MELC have been developed that display altered patterns of response to inducer with respect to terminal cell division (*36, 43–45*). Studies with inhibitors of induced differentiation have been particularly informative. Among these are the tumor promoters, such as 12-*O*-tetradecanoylphorbol-13-acetate (TPA) (*46*, for review), and the glucocorticoid dexamethasone (*47–49*). When MELC are exposed to both HMBA and TPA, the phorbol ester suppresses the onset of terminal cell division as well as the accumulation of globin mRNA and hemoglobin (*46*). When MELC are then transferred from medium containing both inducer and inhibitor into medium without either agent, the cells retain, for a period of time, a ''memory'' of the steps which have occurred during their prior exposure to the inducer in the presence of inhibitor (*50*).

Dexamethasone also suppresses expression of inducer-mediated MELC terminal cell division, as well as inhibiting the accumulation of globin messenger RNAs, globins, and hemoglobins (*51–54*), and we have shown that the glucocorticoid inhibits commitment at a stage in the process of inducer-mediated commitment that is not rate limiting for this process (*54*). Taken together, these studies, as well as other work from our laboratory and others, suggest that there is an early phase in induced MELC differentiation during which certain inducer-mediated metabolic changes occur, including, perhaps, alterations in membrane permeability, cell volume, and cAMP concentration (*1, 11–14, 44, 55*). This is followed by a period during which changes occur that appear to involve the accumulation of a factor or factors that may be responsible for the commitment to terminal cell division and, perhaps, for the expression of genes characteristic of terminal cell differentiation (*54*). This is followed by changes that characterize the expression of the terminal erythroid phenotype (*1, 43, 46*), including synthesis of globin mRNA (*56*), the appearance of heme-synthetic enzymes (*4*), the loss of capacity for cell division (*39, 40*), and other changes (*57–59*).

B. Macromolecular Synthesis and the Expression of Commitment to Terminal Cell Division

Studies involving a number of inhibitors of macromolecular synthesis have provided evidence suggesting that both RNA and protein synthesis are required at different steps in the commitment process (*60, 61*). More recently, studies have also been designed to characterize in detail those steps in HMBA-mediated MELC commitment to terminal cell division that are suppressed by dexamethasone and rapidly expressed upon removal of the steroid. As already noted, MELC exposed to HMBA and dexamethasone show low levels of commitment; upon transfer to medium with inducer but without steroid there is a rapid increase in the proportion of committed cells, within 1–2 hr (*54*). The magnitude of this accelerated expression of commitment is directly proportional to the duration of culture with inducer and steroid, between 12 and 72 hr. Employing inhibitors of macromolecular synthesis during the period of accelerated expression of induced commitment to terminal cell division, Murate and co-workers have demonstrated that this step in the multistep commitment process is dependent upon concurrent protein synthesis but not upon RNA synthesis (*62*). Taken together these observations suggest that HMBA mediates the accumulation of a factor or factors required to prepare MELC to express terminal erythroid cell differentiation and that these factors accumulate during culture with HMBA and the inhibitor, dexamethasone. The evidence is consistent with but does not fully establish that these factors are inducer-mediated mRNA required for the protein synthesis necessary for subsequent expression of commitment. At present, studies are in progress to identify and characterize mRNA species whose synthesis is stimulated by HMBA, even in the presence of dexamethasone. Two candidate mRNA species have, so far, been identified (1.5 and 7.5 kb) by differential screening of cDNA libraries (J. Ravetch, P. A. Marks, and R. Ramsay, unpublished observations).

Little is known of the molecular or biochemical mechanisms that are specifically implicated in the actual implementation of termination of cell division, the putative site of action for the gene products under investigation. The first manifestation of termination of cell division in MELC is a transient prolongation of the G_1 phase of the cell cycle, detected in the first cell cycle that follows a complete S phase in inducer (*63*). After four to five cell divisions induced MELC become permanently arrested in G_1. Evidence has accumulated suggesting that synthesis of a labile protein in early G_1 may be important in the entry of cells into S phase (*64*), and a nuclear protein (p53) has been identified as one candidate for a factor determining normal cell progression from G_1 (or G_0) to S (*65, 66*). Recently, we have shown, by microscopic immunofluorescence, flow microfluorimetry, and immunoprecipitation, that during induced differentiation of MELC there is a decrease in p53 synthesis and in the cell content of this protein

(*67*), mediated, at least in part, by posttranscriptional mechanisms (F. Lovaglio, A. De Leo, F. Real, and R. A. Rifkind, unpublished observations). It is speculated that down-regulation of p53 protein may be part of the coordinated program of events which occurs during induced MELC differentiation and may prove significant for the onset of terminal cell division.

IV. Summary

Inducer-mediated differentiation of murine erythroleukemia cells continues to provide insight into the cellular and molecular mechanisms implicated in cell differentiation. Induced differentiation is accompanied by an acceleration of transcription at the globin loci and, possibly, by posttranscriptional modulation of globin mRNA accumulation as well. Cells at the CFUe-like stage of erythroid cell development represented by the transformed MELC have acquired a unique DNA structure (methylation pattern) and chromatin configuration around the globin genes that distinguishes them from other, nonerythroid cells. Further changes in chromatin configuration accompany and, in part, precede, inducer-mediated acceleration of globin gene transcription. The loss of proliferative capacity appears to reflect a complex multistep process during which the cells accumulate factors (perhaps mRNA) required for the synthesis of proteins that are themselves, in turn, required for expression of the commitment process. The specific biochemical mechanisms implicated in this process remain unknown although a role for down-regulation of p53 protein has been suggested.

References

1. P. A. Marks and R. A. Rifkind, *Annu. Rev. Biochem.* **47,** 419 (1978).
2. M. Terada, L. Cantor, R. A. Rifkind, A. Bank, and P. A. Marks, *Proc. Natl. Acad. Sci. U.S.A.* **69,** 3575 (1972).
3. F. Ramirez, R. Gambino, G. M. Maniatis, R. A. Rifkind, P. A. Marks, and A. Bank, *J. Biol. Chem.* **250,** 6054 (1975).
4. S. Sassa, *in* ''In Vivo and In Vitro Erythropoiesis: The Friend System'' (G. B. Rossi, ed.), p. 219. Elsevier, Amsterdam, 1980.
5. M. Tanaka, J. Levy, M. Terada, R. Breslow, R. A. Rifkind, and P. A. Marks *Proc. Natl. Acad. Sci. U.S.A.* **72,** 1003 (1975).
6. R. C. Reuben, R. L. Wife, R. Breslow, R. A. Rifkind, and P. A. Marks, *Proc. Natl. Acad. Sci. U.S.A.* **73,** 862 (1976).

7. U. Nudel, J. D. Salmon, M. Terada, A. Bank, R. A. Rifkind, and P. A. Marks, *Proc. Natl. Acad. Sci. U.S.A.* **74,** 1100 (1977).
8. U. Nudel, J. Salmon, E. Fibach, M. Terada, R. A. Rifkind, and P. A. Marks, *Cell* **12,** 463 (1977).
9. Y. Ohta, M. Tanaka, M. Terada, O. J. Miller, A. Bank, P. A. Marks, and R. A. Rifkind, *Proc. Natl. Acad. Sci. U.S.A.* **73,** 1232 (1976).
10. V. Volloch and D. Housman, *J. Cell Biol.* **93,** 390 (1982).
11. D. Mager and A. J. Bernstein, *J. Cell. Physiol.* **94,** 275 (1978).
12. D. Mager and A. J. Bernstein, *J. Supramol. Struct.* **8,** 431 (1978).
13. R. L. Smith, I. G. Macara, R. Levenson, D. Housman, and L. Contley, *J. Biol. Chem.* **257,** 773 (1982).
14. Y. Gazitt, R. C. Reuben, A. D. Deitch, P. A. Marks, and R. A. Rifkind, *Cancer Res.* **38,** 3779 (1978).
15. E. Hofer, R. Hofer-Warbinek, and J. E. Darnell, Jr., *Cell* **29,** (1982).
16. M. Sheffery, R. A. Rifkind, and P. A. Marks, *Proc. Natl. Acad. Sci. U.S.A.*, 1180 (1982).
17. H. R. Profous-Juchelka, R. C. Reuben, P. A. Marks, and R. A. Rifkind, *Mol. Cell. Biol.* **3,** 229 (1983).
18. P. Curtis, A. C. Finnigan, and G. Rovera, *J. Biol. Chem.*, **255,** (1980).
19. D. A. Konkel, S. M. Tilghman, and P. Leder, *Cell* **15,** 1125 (1978).
20. Y. Nishioka and P. Leder, *Cell* **18,** 857 (1979).
21. P. Leder, J. Hansen, D. Konkel, A. Leder, Y. Nishioka, and C. Talkington, *Science* **209,** 1336 (1980).
22. A. Leder, D. Swan, F. Ruddle, P. D'Eustachio, and P. Leder, *Nature (London)* **293,** 196 (1981).
23. M. Groudine, M. Peretz, and H. Weintraub, *Mol. Cell. Biol.* **1,** 281 (1981).
24. M. Groudine and H. Weintraub, *Cell* **24,** 393 (1981).
25. H. Weintraub, A. Larsen, and M. Groudine, *Cell* **24,** 333 (1981).
26. G. M. Landes and H. G. Martinson, *J. Biol. Chem.* **257,** 11002 (1982).
27. G. M. Landes, B. Villeponteau, T. M. Pribyl, and H. G. Martinson, *J. Biol. Chem.* **257,** 11008 (1982).
28. B. Villeponteau, G. M. Landes, M. J. Pankvatz, and H. G. Martinson, *J. Biol. Chem.* **257,** 11015 (1982).
29. M. Sheffery, P. A. Marks, and R. A. Rifkind, *J. Mol. Biol.* **172,** 417 (1984).
30. S. C. R. Elgin, *Cell* **27,** 413 (1981).
31. S. Weisbrod, *Nature (London)* **297,** 289 (1982).
32. D. M. Miller, P. Turner, A. W. Neinhuis, D. E. Axelrod, and T. V. Gopulakrishnan, *Cell* **14,** 511 (1978).
33. R. Cohen and M. Sheffery, *J. Mol. Biol.* **182,** 109 (1985).
34. A. S. Tsiftsoglou, W. Wong, V. Volloch, J. Gusella, and D. Housman, *in* "Cell Function and Differentiation, Part A" (G. Akoyunoglou, A. E. Evangelopoulos, J. Georgatsos, G. Palaiologos, A. Trakatellis, and C. P. Tsiganos, eds.), p. 69. Liss, New York (1982).
35. M. Sheffery, R. A. Rifkind, and P. A. Marks, *Proc. Natl. Acad. Sci. U.S.A.* **80,** 3349 (1983).
36. P. A. Marks, Z. X. Chen, J. Banks, and R. A. Rifkind, *Proc. Natl. Acad. Sci. U.S.A.* **80,** 2281 (1983).
37. R. D. Smith, R. L. Seale, and J. Yu, *Proc. Natl. Acad. Sci. U.S.A.* **80,** 5505 (1983).
38. R. D. Smith and J. Yu, *J. Biol. Chem.* **259,** 4609 (1984).
39. J. Gusella, R. Geller, B. Clarke, V. Weeks, and D. Housman, *Cell* **9,** 221 (1976).
40. E. Fibach, R. C. Reuben, R. A. Rifkind, and P. A. Marks, *Cancer Res.* **37,** 440 (1977).

41. R. C. Reuben, R. A. Rifkind, and P. A. Marks, *Biochim. Biophys. Acta* **605,** 325 (1980).
42. P. A. Marks, R. A. Rifkind, A. Bank, M. Terada, R. Gambari, E. Fibach, G. Maniatis, and R. C. Reuben, *in* "Cellular and Molecular Regulation of Hemoglobin Switching" (G. Stamatoyannopoulos and A. W. Neinhuis, eds.), p. 437. Grune & Stratton, New York (1979).
43. P. R. Harrison, *Int. Rev. Biochem.* **15,** 227 (1977).
44. H. Eisen, F. Keppel-Bellivet, C. P. Georgopoulos, S. Sassa, J. Granick, I. Pragnell, and W. Ostertag, *Cold Spring Harbor Conf. Cell Proliferation.* **5,** 277 (1978).
45. I. B. Pragnell, D. J. Arndt-Jovin, T. M. Jovin, B. Fogg, and W, Ostertag, *Exp. Cell Res.* **125,** 459 (1980).
46. P. A. Marks, R. A. Rifkind, R. Gambari, E. Epner, Z. X. Chen, and J. Banks, *Curr. Top. Cell. Regul.* **21,** 198 (1982).
47. M. G. Santoro, A. Benedetto, and B. B. Jaffe, *Biochem. Biophys. Res. Commun.* **85,** 1510 (1978).
48. A. S. Tsiftsoglou, J. F. Gusella, V. Volloch, and D. Housman, *Cancer Res.* **39,** 3849 (1979).
49. H. B. Osborne, A. C. Bakke, and J. Yr, *Cancer Res.* **42,** 513 (1982).
50. E. Fibach, R. Gambari, P. A. Shaw, G. Maniatis, R. C. Reuben, S. Sassa, R. A. Rifkind, and P. A. Marks, *Proc. Natl. Acad. Sci. U.S.A.* **76,** 1906 (1979).
51. S. C. Lo, R. Aft, J. Ross, and G. C. Mueller, *Cell* **15,** 447 (1978).
52. W. Scher, D. Tsuei, S. Sassa, P. Price, N. Gabelmar, and C. Friend, *Proc. Natl. Acad. Sci. U.S.A.* **75,** 2851 (1978).
53. R. C. Mierendorf and G. C. Mueller, *J. Biol. Chem.* **256,** 6736 (1981).
54. Z. X. Chen, J. Banks, R. A. Rifkind, and P. A. Marks, *Proc. Natl. Acad. Sci. U.S.A.* **79,** 471 (1982).
55. Y. Gazitt, A. D. Deitch, P. A. Marks, and R. A. Rifkind, *Exp. Cell Res.* **117,** 413 (1978).
56. J. Ross, Y. Ikawa, and P. Leder, *Proc. Natl. Acad. Sci. U.S.A.* **69,** 3620 (1972).
57. S. H. Boyer, K. D. Wuu, A. N. Noyes, R. Young, W. Scher, C. Friend, H. Preisler, and A. Bank, *Blood* **40,** 823 (1972).
58. W. Ostertag, H. Melderis, G. Steinheider, N. Kluge, and S. Dube, *Nature (London), New Biol.* **239,** 231 (1972).
59. H. Eisen, S. Nasi, C. P. Georgopoulos, D. Arndt-Jovin, and W. Ostertag, *Cell* **10,** 689 (1977).
60. R. Levenson, J. Kerner, and D. Housman, *Cell* **18,** 1073 (1979).
61. R. Levenson and D. Housman, *J. Cell Biol.* **82,** 715 (1979).
62. T. Murate, T. Kaneda, R. A. Rifkind, and P. A. Marks, *Proc. Natl. Acad. Sci. U.S.A.* **81,** 3394 (1984).
63. M. Terada, J. Fried, U. Nudel, R. A. Rifkind, and P. A. Marks, *Proc. Natl. Acad. Sci. U.S.A.* **74,** 248 (1977).
64. P. W. Rossow, V. G. H. Riddle, and A. B. Pardee, *Proc. Natl. Acad. Sci. U.S.A.* **76,** 4446 (1979).
65. J. Milner and S. Milner, *Virology* **112,** 785 (1981).
66. W. E. Mercer, D. Nelson, A. B. DeLeo, L. J. Old, and R. Baserga, *Proc. Natl. Acad. Sci. U.S.A.* **79,** 6309 (1982).
67. D.-W. Shen, F. X. Real, A. B. DeLeo, L. J. Old, P. A. Marks, and R. A. Rifkind, *Proc. Natl. Acad. Sci. U.S.A.* **80,** 5919 (1983).

7

Notes on Tissue-Specific Gene Control*

J. E. DARNELL, JR., D. F. CLAYTON, J. M. FRIEDMAN, AND D. J. POWELL

The Rockefeller University
New York, New York

When mobile elements were becoming widely appreciated about 10 years ago (*1, 2*) and DNA rearrangements of the immunoglobin genes in lymphocytes were definitely documented (*3;* reviewed in Ref. *2*), there was a rush to judgment by many molecular biologists that at least determination if not differentiation might often be based on changes in primary DNA sequences. Subsequent work has not confirmed this suspicion, and emphasis has shifted back to differential transcriptional controls as the basis for differentiated gene function during development (*4*). In addition, differential RNA processing for a growing number of primary RNA products has been found not only in viruses but also in both invertebrate and mammalian genes: hormone production, for example (reviewed in Ref. *4*).

It is still too early to suggest that DNA rearrangements will play a role only in the development of cells of the immune system. For example, it is popular to suppose that the central nervous system might employ DNA rearrangements in achieving the almost endless diversity of function of nerve cells that is needed to construct and operate a nervous system. One accompaniment to the use of random DNA rearrangements within the immunoglobulin V and C genomic regions is frequent error and loss of cells that do not perform well (*2*). In the

*This work was supported by grants from the National Institutes of Health (CA 16006-11 and CA 18213) and the U.S. National Cancer Institute (CA 123M).

GENETICS, CELL DIFFERENTIATION, AND CANCER

nervous system also there is considerable trial and error during CNS development, encouraging views that rearrangements might be important here. Nevertheless the overwhelming suggestion at present is that differential transcriptional control is a mainstay of both determination and differentiation (*4*). In fact one view of development suggests a series of gene activations beginning in the fertilized egg that lead inevitably to a cascade of additional gene activations. Some of these events would occur early and activate gene products such as receptor proteins and DNA binding proteins, and these would be recognized as determinative steps in development. These would be the decisions that lead cells into different pathways. Later events would activate gene products that we recognize in specialized tissues, and this would be called differentiation. In this view of determination and differentiation the gene activations at the molecular level could be the same, for example, positive-acting factors that bind to DNA to alter chromatin structure or directly promote attachment of RNA polymerase. The only difference in determination and differentiation, therefore, would be which gene activation came first.

If gene activations at the transcriptional level are an important principle underlying development, one of the central features in any overall developmental plan must be some mechanism that triggers the coordinated expression of genes that are all active in a particular tissue or a particular cell type. This coordination of production of tissue-specific proteins is nowhere as striking as in the mammalian liver, which has been singled out by classical biochemistry as containing and secreting perhaps the largest variety of tissue-specific, in fact cell (hepatocyte)-specific, products. To gain some insight about coordinated gene expression and its relationship to developmental decisions in general, my colleagues and I have been studying mouse liver for several years, and in this article we summarize some conclusions and ideas that arise from these experiments.

In order to measure liver-specific mRNA biosynthesis, a series of cDNA clones complementary to mRNA found in the liver but not complementary to that found in brain were selected (*5*). The range of concentration of specific-sized mRNAs complementary to these cDNA clones was also tested in brain, kidney, spleen, and intestinal tissue as well as in the liver (*6*). Of the approximately 12 mRNA sequences tested in all tissues 3 were found only in the liver and 7 were present at approximately 50 to 100 times the concentration in liver as in other tissues; 2 were not present in spleen or brain but were in both liver and kidney at comparable concentrations. To determine the level at which this differential gene expression was controlled, isolated nuclei were prepared from tissues (liver, kidney, brain, and spleen) and allowed to elongate already initiated RNA chains in the presence of [α-^{32}P]UTP. Such reactions in nuclei from cultured cells produce a profile of ^{32}P-labeled RNA complementary to various cDNAs that is indistinguishable from the profile of labeled RNA obtained with brief [^{3}H]uridine pulses of whole cells (*5, 7, 8*). This we take as confirmation that differential rates of RNA synthesis from different genes is accurately assessed by the analysis of such labeled nascent RNA from isolated nuclei.

When the labeled RNAs from liver, brain, kidney, and spleen were examined a clear result emerged (*5, 6*). Transcriptional control was the major but not the sole basis for the presence of the specific mRNAs at high concentrations in the liver. Transcriptional signals for 10 of the 12 liver-specific mRNAs were found only in liver nuclei. In the two cases in which mRNAs were shared between kidney and liver (the cDNAs labeled 10 in Ref. *6*) there was a substantial transcriptional signal in tissue other than the liver. Here the transcriptional signal was definitely present in the kidney at a lesser intensity than in liver but the mRNA contents of the tissues were similar, suggesting a substantially greater degree (approximately three- to fivefold) of stability of kidney mRNAs compared to liver mRNAs.

This differential transcription by polymerase II of specific genes in the liver was not matched by differential transcription of tubulin, actin, tRNAs, or rRNAs. These were similar in all nuclei. However, the cytoplasmic concentration of the actin and tubulin sequences were substantially different between tissues (*6*). This again points to a strong posttranscriptional bias, conserving, for example, the actin and tubulin mRNA sequences in spleen and brain compared to liver. A similar result with increased preservation of actin and tubulin mRNAs was found in regenerating liver (*9*). Here the amounts of actin and tubulin mRNAs increased by as much as 15-fold during the 48 hr of regeneration but transcription went up only threefold during 6 hr or less for actin and not at all for tubulin.

A general rule might be suggested from these studies. The overall balance needed to achieve a tissue-specific distribution of proteins is achieved mainly at the transcriptional level for tissue-specific proteins and is achieved at the posttranscriptional level at least for some proteins that are common to many tissues. In addition, changing levels of proteins in growing compared to resting cells of the same type may often be accomplished by posttranscriptional changes.

We will summarize two further aspects of our work on transcriptional control of liver-specific mRNA sequences. In these experiments we draw attention to two aspects of transcriptional control: (1) what is the signal(s) responsible for instigating and maintaining a tissue-specific transcriptional pattern, and (2) what is the meaning of a low level of transcription of a tissue-specific pattern compared to a high level of tissue-specific transcription? Methodologically, the experiments concern a comparison of transcription rates and mRNA concentrations in liver cells fresh from an animal and in the same cells cultured for a day either as dispersed cells (*10, 11*) or as cultured tissue fragments. A second comparison has also been made between liver cells and hepatoma cells for the same two parameters (*12*).

Let us first consider the effect of disaggregation of liver cells on the transcriptional rate of a series of liver-specific mRNAs. Hepatocytes can be dispersed by either a protease treatment infused through the hepatic circulation or a treatment with chelating agents, the major factor in both cases being a release of desmosomal connections between adjacent cells (*14*). Such cells continue to transcribe liver-specific mRNA sequences when first isolated but gradually, during the first 24 hr outside the body, the cells undergo a specific loss of transcriptional activity for the tissue-specific sequences (*10*). Other RNA polymerase II transcripts either increase briefly (actin, for example) or return to near-normal levels and remain more or less constant thereafter for several days in culture. The eventual result in the decrease of liver-specific mRNA transcription is the loss of liver-specific mRNAs in the cytoplasm when the liver cells are continued in culture. This is the result of normal mRNA turnover, we presume. The loss of mRNA in culture can be slowed substantially for the liver-specific mRNAs by culturing the cells without serum. Thus it appears that a stabilization of the liver-specific mRNAs in the absence of their synthesis can be modulated by environmental changes (*15*).

This dramatic drop in tissue-specific transcription is not observed, at least not nearly to the same extent, if the liver is chopped to fine tissue segments (about 10 cells on a side) and cultured in the same medium (*11*). In this experiment, however, the cells are maintained in the same tissue configuration as in the animal. The major basis for the loss of the specific transcriptional signal in the dispersed cells is presumably not due to loss of nutritional (circulating) compounds since the medium is the same for dispersed cells and tissue chunks. Rather, the decreased transcription would appear to be related to the loss of contact or a change in tissue organization or a change in cell shape. While the transcriptional rate for the tissue-specific genes declines, it does not completely cease. If autoradiograms are exposed for longer times, the tissue-specific pattern of transcription can still be observed. The decline is perhaps between 5- and 50-fold for the various liver-specific transcripts. Thus the loss is one of rate and not of specificity.

These results from differentiated liver cells were compared next to those obtained in hepatoma cells (*12*). Hepatomas that retain the ability to synthesize liver-specific secretory proteins and enzymes have been well recognized for some time (*16*). In some of these tumor cells, liver function is quite high. For example, albumin production is stated to be near normal in some hepatomas. We examined the levels of mRNAs with the rat hepatoma line (termed FAO) and found liver-specific mRNA concentrations were ~15–100% that found in liver [albumin mRNA was 15%, phosphoenol pyruvate carboxy kinase (PEPCK) mRNA was 100%]. However, the transcriptional rates of liver-specific mRNAs in hepatomas was low, ~4% for albumin and 10% for PEPCK and for transferrin, compared to the liver. Again, as in the primary cells cultured without serum, more liver-specific mRNA was conserved in the cultured cell than in the liver itself, and transcription was decreased by a factor of 10 or more in the cultured cells compared to the liver nucleus.

From these various results we make several conclusions. The maximal rate of tissue-specific transcription may require tissue organization or at least correct intracellular organization. This effect of cell–cell contact or cell shape is a large one, accounting for >90% of the specific transcription output in adult liver cell nuclei. Perhaps there is an architectural counterpart to the differentiated state that must be maintained for maximal rates of transcription (*17*). Fetal cells transcribe less rapidly many of the liver-specific sequences than does the adult liver (*6*). During late fetal life the hepatocyte is known to undergo the formation of tight junctions (*18*), and perhaps there is a an effect on transcriptional rate of such morphologic events. At any rate fetal cells, dispersed primary hepatocytes, and hepatomas all transcribe many of the mRNA sequences found specifically in adult liver at a substantially decreased rate. It appears possible, therefore, that two levels of activation for the entire range of liver-specific transcription might operate. An initial step would be responsible for differential activation of liver-specific RNA synthesis. A second step would greatly augment the specifically activated genes.

A final point concerns the use of cell cultures to examine newly introduced genes that are subject to tissue-specific transcriptional control. This popular and necessary approach is currently in wide use. We can anticipate learning a great deal from such studies about transcriptional factors that are tissue specific. However, to understand finally the differentiated state we have to be concerned with the rates of transcription, which are, at least for liver, very different in true liver tissue and dispersed cells. The factors responsible for these large differences may not be immediately discovered with the present very valuable techniques of gene purification and transfection.

References

1. A. I. Bukhari, J. A. Shapiro, and S. L. Adhya, (eds.). "DNA Insertion Elements, Plasmids and Episomes." Cold Spring Harbor Laboratory, New York, 1977.
2. "Movable Genetic Elements," Vol. 45. Cold Spring Harbor Laboratory, New York, 1980.
3. N. Hozumi and S. Tonegawa. *Proc. Natl. Acad. Sci. U.S.A.* **73,** 3628 (1976).
4. J. E. Darnell, *Nature (London)* **297,** 365 (1982).
5. E. Derman, K. Krauter, L. Walling, C. Weinberger, M. Ray, and J. E. Darnell, *Cell* **23,** 73 (1981).
6. D. J. Powell, J. M. Friedman, A. J. Oulette, K. S. Krauter, and J. E. Darnell, *J. Mol. Biol.* **179,** 21–35 (1984).
7. E. Hofer and J. E. Darnell, *Cell* **23,** 585 (1981).
8. J. E. Darnell, Jr., M. Salditt-Georgieff, D. F. Clayton, K. S. Krauter, B. A. Citron, D. J. Powell, and E. Hofer, *in* "Transfer and Expression of Eukaryotic Genes" (H. Vogel, ed.). Academic Press, New York (1984).
9. J. M. Friedman, and J. E. Darnell, *J. Mol. Biol.* **179,** 37–53 (1984).
10. D. F. Clayton and J. E. Darnell, *Mol. Cell. Biol.* **3,** 1552 (1983).
11. D. F. Clayton, A. Harrelson, and J. E. Darnell, *Mol. Cell. Biol.* submitted (1985).
12. D. F. Clayton, M. C. Weiss, and J. E. Darnell, *Mol. Cell. Biol.* submitted (1985).
14. H. C. Pitot, and A. E. Sirica, *Methods Cell Biol.* **21B,** 441 (1980).
15. D. C. Jefferson, D. F. Clayton, J. E. Darnell, and L. M. Reid, *Mol. Cell. Biol.* **4,** 1929 (1984).
16. J. Deschatrette, E. E. Moore, M. Dubois, and M. C. Weiss, *Cell* **19,** 1043 (1980).
17. D. Mathog, M. Hochstrasser, Y. Gruenbaum, H. Saumweber, and J. Sedat, *Nature (London)* **308,** 414 (1984).
18. R. Montesano, D. S. Friend, A. Perrelet, and L. Orci, *J. Cell Biol.* **67,** 310 (1975).

PART II

Cancer Genes and Viruses

Activation of *c-myc* in Viral and Nonviral Neoplasia

K. WIMAN, B. CLURMAN, C.-K. SHIH,* M. GOODENOW, M. SIMON,* R. LESTRANGE, W. HAYWARD

Molecular Biology and Virology Program of the Graduate School
Memorial Sloan-Kettering Cancer Center
New York, New York

A. HAYDAY, AND S. TONEGAWA

Massachusetts Institute of Technology
Cambridge, Massachusetts

I. Introduction

The *c-myc* gene has been implicated in a wide variety of neoplasms, including lymphomas, leukemias, and carcinomas (*1–3*). Although the oncogenic potential of this gene can be activated by such diverse mechanisms as insertion of retroviral regulatory sequences, translocation, and gene amplification (*4–12*), it is perhaps significant that in each case the mutational event induces a change in gene expression. In this chapter we review our recent efforts to elucidate the

*Also The Rockefeller University, New York, New York.

mutational events responsible for *c-myc* activation in two biological systems: ALV-induced B-cell lymphomas and human Burkitt lymphomas.

II. ALV-Induced B-Cell Lymphomas

Induction of B-cell lymphomas by avian leukosis virus (ALV) results from insertion of proviral sequences adjacent to, and transcriptional activation of, the host *c-myc* gene (*4*). This conclusion is based on the following lines of evidence:

1. In more than 80% of ALV-induced lymphomas proviral sequences are integrated adjacent to the *c-myc* gene.
2. Levels of *c-myc* mRNA are elevated 30- to 100-fold in the lymphomas, compared to equivalent normal tissues.
3. In most tumors, the *myc*-specific mRNA contains viral sequences derived from the long terminal repeat (LTR) of the integrated provirus (for a review see Ref. *1*).

Thus, transcription initiates within the viral LTR and reads into the adjacent *c-myc* gene, causing transcriptional activation of the gene.

In approximately 70% of ALV-induced lymphomas, proviral insertions are located within a short (270-bp) region in the first intron of *c-myc* (Fig. 1) (*13*). Less frequently, integration occurs within the first exon, which is noncoding (Fig. 1), or upstream of the cap site (*14*). One example of a downstream integration has also been reported (*5*). This nonrandom distribution of integration sites may result from preferential integration within certain DNA sequences or chromatin structures. Alternatively, it may reflect a selection for integration events that activate the gene most efficiently. No integrations have been found within the coding region, suggesting that there is a strong selective pressure to maintain the integrity of the *c-myc* protein.

III. Human Burkitt Lymphomas

In the majority of Burkitt lymphomas, the *c-myc* gene is translocated from its normal position on chromosome 8(q24) (*9, 15*) and joined to the *IgH* locus on chromosome 14 (*8, 9*). In the remaining minority of cases *c-myc* is joined to one of the light chain loci. A similar translocation event has been observed in murine plasmacytomas (*7, 8*).

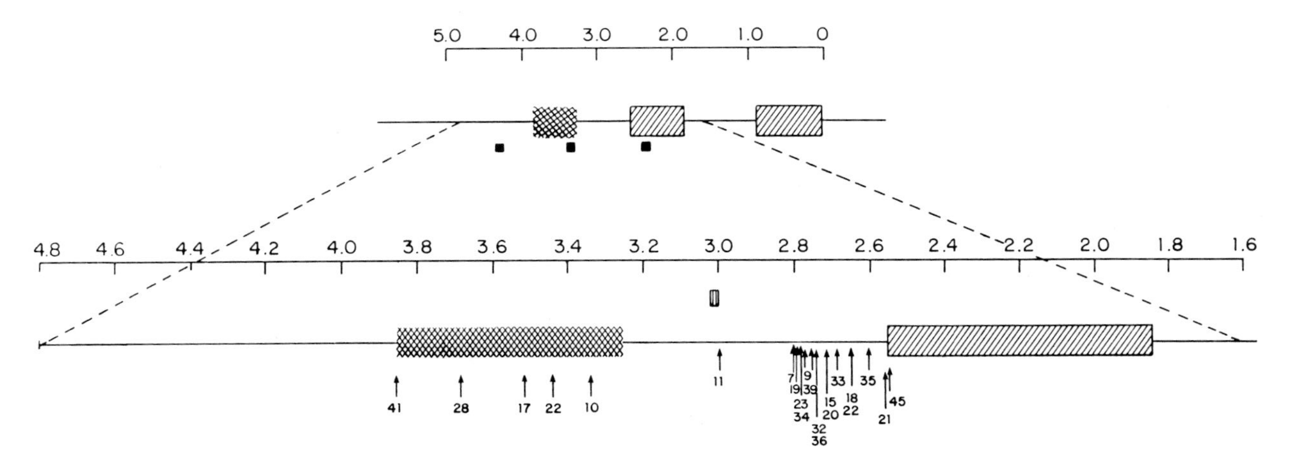

Fig. 1. Sites of proviral integrations in avian B-cell lymphomas. The two *c-myc* coding exons are designated by single hatched boxes; 12 additional nucleotides, present at the 5′ end of MC29 *v-myc*, are indicated above the enlarged map at 3.0 kb. The precise boundaries of the first exon have not been firmly established. The hatched box shows the location of the restriction fragment used as probe to detect the first exon. A sequence within the second exon, and two upstream regions that demonstrate complementarity to this sequence, are designated by black boxes beneath the upper *c-myc* map. From Ref. *13.*

Efforts to deduce the mechanism by which translocation causes activation of *c-myc* have been complicated by the fact that the juxtaposition of sequences from the two loci is variable from one tumor to another (*7–9, 16–21*). In a high proportion of murine plasmacytomas and in many Burkitt lymphomas, the breakpoint in *c-myc* is located within the first intron, but breakpoints within, or considerably upstream of, the first exon are also common. Breakpoints in the *Ig* locus often occur within the switch region, but sites have also been observed both upstream and downstream of this region.

Two examples of *c-myc* rearrangements are shown in Fig. 2. In the Manca cell line, the breakpoints are within the first intron of *c-myc* and within the intron upstream of Sμ in the *IgH* locus (*16, 17*). The resulting rearranged *c-myc* gene thus lacks the first exon and carries an *IgH* enhancer sequence just upstream of the *c-myc* gene (*17*).

This is perhaps the easiest case to explain, as the *Ig* enhancer would be expected to induce elevated and constitutive expression of the translocated *c-myc* gene. Several other examples of this type of rearrangement have recently been

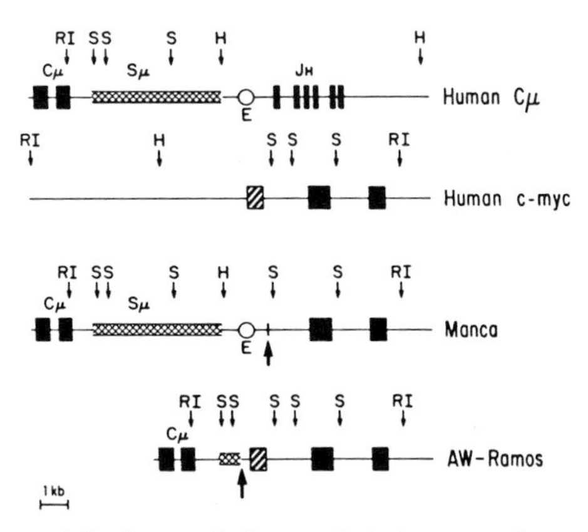

Fig. 2. Maps of part of the human *Ig* heavy chain locus on chromosome 14 (Cμ), the normal human *c-myc* gene on chromosome 8, and the translocated *c-myc* genes of the Manca and AW-Ramos B-cell lymphoma lines. The heavy chain constant domain (Cμ), the switch region (Sμ), and the six functional human J segments (J_H) of the *Ig* locus are indicated. E denotes the position of a recently identified transcriptional enhancer element (*17*). The first, noncoding exon of the human *c-myc* gene is represented by the hatched box; black boxes indicate coding exons. Large arrows indicate chromosomal junction points in Manca and AW-Ramos. Restriction enzyme cleavage sites are *Eco*RI (RI), *Hin*dIII (H), and *Sac*I (S). Transcription of the *Ig* locus is from right to left; transcription of *c-myc* is from left to right. From Ref. *21*.

described (*22, 23*). In most cases, however, the enhancer is not linked to *c-myc*. In the AW-Ramos lymphoma line, for example (Fig. 2), recombination has occurred within the μ switch region, downstream of the enhancer (*21*). The point of chromosomal junction at the *c-myc* locus is 340 bases upstream of the 5′ cap site, leaving all three exons intact. Nucleotide sequencing demonstrates that there are no mutations within the coding sequences of the rearranged *c-myc* gene in AW-Ramos. Furthermore, the mRNA structure appears to be normal. Thus, altered properties of the *c-myc* gene product cannot account for activation of this gene. Two base substitutions were found in noncoding regions of the *c-myc* gene (in the first exon, and in the region between the cap site and the chromosomal recombination point).

The data thus indicate that altered expression, rather than an altered gene product, is responsible for *c-myc* activation in AW-Ramos cells, and that this is a consequence of either the loss of regulatory sequences located more than 340 nucleotides upstream of *c-myc,* or disruption of normal *c-myc* regulation by one or both base substitutions. Alternatively, the presence of as yet unidentified enhancerlike sequences in the *Ig* locus may cause altered expression of *c-myc*.

IV. Identification of *cis*-Acting Transcriptional Signals

The common feature of all *c-myc* activations, regardless of the mechanism involved, or the mutagenic agent responsible for the alterations, is a disruption of the normal regulated expression of the gene. For this reason, we have initiated studies designed to identify and characterize the regulatory sequences responsible for control of *c-myc* expression in the normal cell.

We have linked upstream sequences of both chicken and human *c-myc* genes to the chloramphenicol acetyltransferase gene (*CAT*). Constructs containing different lengths of upstream sequences from the chicken *c-myc* gene were introduced into chicken embryo fibroblasts (CEF) by the calcium phosphate precipitation technique (*24*). Likewise, CAT constructs carrying upstream sequences from the human *c-myc* gene were introduced into NIH3T3 and HeLa cells (*25*). These studies demonstrated that the region immediately upstream of the first exon can promote transcription of the *CAT* gene. Further upstream are sequences that appear to exert a negative effect on expression, but we are interpreting these observations cautiously because of the inherent difficulties and potential artifacts in the assay system.

V. Conclusions

Many different types of mutations have been identified in activated *c-myc* genes (*1–12, 16–21*), including alterations within coding sequences that would result in the generation of an altered *c-myc* protein (*19, 26*). In at least two cases, however, the nucleotide sequence of the coding region of the activated *c-myc* gene has been shown to be identical to that of the normal gene (*18, 21*). It seems clear, therefore, that mutations that result in an altered gene product are not an essential feature of *c-myc* activation. By contrast, all cases of *c-myc* activation reported thus far appear to involve a perturbation in the normal control of *c-myc* expression. It is hoped that an understanding of the mechanisms involved in regulation of the *c-myc* gene in the normal cell, and of the molecular changes that disrupt this control in the malignant cell, will lead to new approaches to overcome the consequences of these mutational events in *c-myc*–related human cancers.

Acknowledgments

The authors would like to thank Lauren O'Connor for help in preparation of the manuscript. Studies conducted in W.H.'s laboratory were supported by NIH grant CA 34502, the Flora E. Griffin Fund, The Kleberg Foundation, and Kimberly-Clark. K.W. is the recipient of an EMBO fellowship. M.G. is the recipient of a Damon Runyon/Walter Winchell Cancer Fund Fellowship.

References

1. W. S. Hayward, B. G. Neel, and S. M. Astrin, *Adv. Viral Oncol.* **1,** (1982).
2. J. M. Bishop, *Ann. Rev. Biochem.* **52,** 301–354 (1983).
3. J. J. Yunis, *Science* **221,** 227 (1983).
4. W. S. Hayward, B. G. Neel, and S. M. Astrin, *Nature (London)* **291,** 475 (1981).
5. G. S. Payne, J. M. Bishop, and H. E. Varmus, *Nature (London)* **295,** 209 (1982).
6. Y.-K. T. Fung, L. B. Crittenden, and H.-J. Kung, *J. Virol* **44,** 742 (1982).
7. G. L. C. Shen-Ong, E. J. Keath, S. P. Piccoli, and M. D. Cole, *Cell* **31,** 443 (1982).
8. R. Taub, I. Kirsch, C. Morton, G. Lenoir, D. Swan, S. Tronick, S. Aaronson, and P. Leder, *Proc. Natl. Acad. Sci. U.S.A.* **79,** 7837 (1982).
9. R. Dalla-Favera, M. Bregni, J. Erickson, D. Patterson, R. Gallo, and C. M. Croce, *Proc. Natl. Acad. Sci. U.S.A.* **79,** 7824 (1982).

10. S. Collins, and M. Groudine, *Nature (London)* **298,** 679 (1982).
11. K. Alitalo, M. Schwab, C. C. Lin, H. E. Varmus, and J. M. Bishop, *Proc. Natl. Acad. Sci. U.S.A.* **80,** 1707 (1983).
12. C. D. Little, M. M. Nau, D. N. Carney, A. F. Gazdar, and J. D. Minna, *Nature (London)* **316,** 194 (1983).
13. C.-K. Shih, M. Linial, M. M. Goodenow, and W. S. Hayward, *Proc. Natl. Acad. Sci. U.S.A.* **81,** 4697 (1984).
14. H. Robinson, personal communication.
15. B. G. Neel, S. C. Jhanwar, R. S. K. Chaganti, and W. S. Hayward, *Proc. Natl. Acad. Sci. U.S.A.* **79,** 7842 (1982).
16. H. Saito, A. Hayday, K. Wiman, W. S. Hayward, and S. Tonegawa, *Proc. Natl. Acad. Sci. U.S.A.* **80,** 7476–7480 (1983).
17. A. C. Hayday, S. D. Gillies, J. Saito, C. Wood, K. Wiman, W. Hayward, and S. Tonegawa, *Nature (London)* **317,** 334 (1984).
18. J. Battey, C. Moulding, R. Taub, W. Murphy, T. Steward, H. Potter, G. Lenoir, and P. Leder, *Cell* **34,** 779 (1983).
19. T. H. Rabbitts, P. H. Hamlyn, and R. Baer, *Nature (London)* **316,** 760 (1984).
20. K. B. Marcu, L. J. Harris, L. W. Stanton, J. Erickson, R. Watt, and C. M. Croce, *Proc. Natl. Acad. Sci. U.S.A.* **80,** 519 (1983).
21. K. G. Wiman, B. Clarkson, A. C. Hayday, H. Saito, S. Tonegawa, and W. S. Hayward, *Proc. Natl. Acad. Sci. U.S.A.* **81,** 6798 (1984).
22. U. Siebenlist, L. Hennighausen, J. Battey, and P. Leder, *Cell* **37,** 381 (1984).
23. M. Diaz, personal communication.
24. B. Clurman, *et al.,* unpublished.
25. K. Wiman, *et al.,* unpublished.
26. D. Westaway, G. Payne, and H. E. Varmus, *Proc. Natl. Acad. Sci. U.S.A.* **80,** 843 (1983).

Retroviruses and Cancer Genes

J. MICHAEL BISHOP

Department of Microbiology and Immunology
and The George Williams Hooper Foundation
University of California Medical Center
San Francisco, California

I. Introduction

The oncogenes of retroviruses arose by transduction of protooncogenes found in the genomes of vertebrates and other metazoan organisms (*1*). Three large puzzles arise from this scheme. First, by what means do viral oncogenes act? How do they evoke and sustain the neoplastic phenotype? Second, the conservation of protooncogenes over immense lengths of evolutionary time implies that these genes serve vital roles in the economy of the normal cell and organism. What might those roles be? Third, the tumorigenicity of retroviral oncogenes raises the possibility that their cellular parents might be our own ''cancer genes'' in thin disguise, the substrates for carcinogens of various sorts. I review here experimental strategies and findings that bear on each of these puzzles.

II. Retroviral Oncogenes

Almost two dozen oncogenes have now been found in retroviruses (Table I) (*2*). Although enzymatic functions have been assigned to the products of some of these genes (tyrosine-specific protein kinases are the most prevalent and extensively explored) and others have at least physiological identities (a derivative of the receptor for epidermal growth factor and a subunit of platelet growth factor), we can in no instance explain how the action of these genes elicits neoplastic growth. I will illustrate some of the qualities of this conundrum with a brief description of two apparently related oncogenes, *v-myb* and *v-myc*.

Both *v-myb* and *v-myc* can transform monocytic hemopoietic cells, and it was once thought that transformation resulted from arrest of cellular differentiation. It is now apparent, however, that this scheme is not likely to be correct. First, both

TABLE I

Products of Retroviral Oncogenes[a]

Tumorigenicity	Oncogene	Properties of product	
Sarcomas	*src*	Tyr-specific protein kinase on/in plasma membrane (*src*, *yes*, *fps/fes*, *ros*, *fgr*, *fms*, *abl*)	Kinase gene family (*src* through *mos*)
	yes		
	fps/fes		
	ros		
	fgr		
	fms		
B-Lymphoma	*abl*		
Erythroleukemias	*erb-B*	Membrane glycoprotein: truncated egf receptor	
Sarcomas	*raf/mil*	Ser/Thr protein kinase	
Sarcomas	*mos*	Phosphoprotein in cytosol	
Erythroleukemias and sarcomas	*ras*	GTP-binding protein/GTPase on/in plasma membrane	
Carcinomas, sarcomas, and myelocytic leukemia	*myc*	Nuclear proteins (*myc*, *myb*, *fos*, *ski*)	
Myeloblastic leukemia	*myb*		
Osteosarcomas	*fos*		
Sarcomas	*ski*		
Sarcomas	*sis*	Cytoplasmic homolog of PDGF	

[a] Listed are the retroviral oncogenes whose protein products have been identified and at least provisionally characterized. The kinase gene family is defined by similarities between the amino acid sequences of the proteins encoded by the genes, not necessarily by shared functions.

v-myb and *v-myc* elicit neoplastic growth following introduction of the genes into either the primitive stem cell or the mature product (the macrophage) of the developmental lineage (*13*). Second, the phenotypes of cells transformed by either *v-myb* or *v-myc* cannot be construed as representing single compartments within the developmental lineage (Table II) (see also Ref. *3*). Third, the action of each oncogene is epistatic over the other for certain phenotypic properties but not for others (Table II). The actions of *v-myb* and *v-myc* therefore evoke abnormal phenotypes that cannot be readily explained simply by arrest of differentiation.

The products of *v-myb* and *v-myc* may perform related functions:

1. There are tenuous but nevertheless perceptible relationships between the structures of the proteins (*4*).
2. The products of both genes reside in the nuclear matrix of transformed cells (G. Ramsay, unpublished observations) and bind to double-stranded DNA (*5;* K.-H. Klempnauer, unpublished observations).
3. Both genes supplement the action of another oncogene (a mutant version of *c-ras*) in the transformation of primary cultures of rat embryo cells (*6;* and L. Parada, personal communication).

The only clue as to how these genes might function, however, is a provisional suggestion that the product of *v-myc* augments the activity of transcriptional promoters for diverse sorts of genes (R. Kingston, personal communication).

The contrast between *v-myb* and the transformed cell is not irrevocable. Phor-

TABLE II

The Separate and Combined Effects of *v-myb* and *v-myc* in the Myelomonocytic Hemopoietic Lineage[a]

	Blast	*v-myb*	*v-myb* + *v-myc*	*v-myc*	Macrophage
Cell size	+	+	++	+++/++++	++++
Adherence	−	−	−	+/++	++++
Fc-receptor	−	++	++	++++	++++
Immune phagocytosis	−	−	+/++	++++	++++
C3-receptor	−	−	+/++	+/++	+++/++++
Phorbol-induced differentiation	NT	Slow	Rapid	Rapid	NT

[a] Phenotypic properties of normal and transformed cells were evaluated as detailed in Ref. 6 and graded with a scale ranging from undetectable (−) to maximum (++++). NT denotes not tested.

bol esters cause cells transformed by *v-myb* to differentiate to macrophages; the tumorigenic phenotype is reversed (*7*). The production of *v-myb* protein continues unabated as the cells differentiate (*7*), but the protein relocates from the nucleus to a cytoplasmic organelle that is probably the Golgi apparatus (*8*). We cannot at present distinguish cause from effect in these events, but we can perceive prospects for insight into both the mechanisms by which hemopoietic differentiation is controlled and the signals that direct proteins to various compartments within the cell.

III. Protooncogenes in *Drosophila melanogaster*

Most observers suspect that protooncogenes are components of the programs that control growth and development. In search of genetic purchase on this hypothesis, we have exploited the discovery that *Drosophila melanogaster* possesses at least some of the known protooncogenes (*9*). We first sought genes related to *v-src*. Three were found, located at chromosomal positions 64B, 73B, and 29A (*10*). The first of these is apparently analogous to *c-src* in mammalian and avian cells, the second to *c-abl,* and the third remains unidentified. How are these genes used? What is the purpose of the genetic diversification that they represent? Might they have been fabricated for different developmental programs?

As a first step toward answering these questions, we analyzed the expression of the three genes during the course of *Drosophila* development. The genes are expressed in distinctive patterns (Table III), as if they were used at different times and for different purposes during the course of development (M. Simon, unpublished observations). Moreover, *c-src*(*Drosophila*) gives rise to three RNAs of different sizes, and each of these is in turn produced at different times during development. The impressions gained from these results are now being tested by mutagenesis directed at *c-src*(*Drosophila*).

The map of protooncogenes in *Drosophila melanogaster* has become quite "busy" (Fig. 1). Included are several analogs of mammalian *c-ras* and a newly discovered analogue of *c-myb* at position 13F on the X chromosome—the first sighting of a *Drosophila* protooncogene that might encode a nuclear protein (A. Katzen, unpublished observations). Each of these loci deserves dissection by genetic strategies: there is no system more tractable for these purposes than *Drosophila!*

TABLE III

Expression of Protooncogenes during the Development of *Drosophila melanogaster*[a]

RNA	Size (kb)	Periods of relative abundance
64B (*src*)	3.5	2–3 hr
64B	5.0	2–9 hr
64B	5.5	3–12 hr/pupal
73B (*abl*)	6.0	0–1 hr/pupal
29A (?)	3.5	0–21 hr/larval/pupal

[a] RNAs transcribed from the designated *Drosophila* genes were fractionated by electrophoresis in agarose gels and detected by molecular hybridization. The *c-src* gene at position 64B gives rise to three mRNAs with the sizes given. The other genes are each represented by single mRNAs.

IV. Protooncogenes and Cancer

Several forms of genetic damage are known to afflict protooncogenes in human cancers (reviewed in Ref. *11*). Point mutations change the function of the gene product, chromosomal translocations can alter expression of the affected

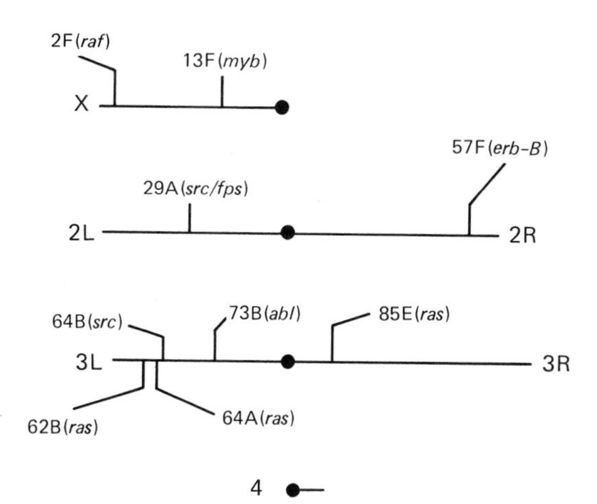

Fig. 1. Protooncogenes in *Drosophila melanogaster.* The genes are identified according to their positions on *Drosophila* chromosomes and the protooncogenes to which they appear to be most closely related. The gene at 29A is related to both *c-src* and *c-fps* but has not been characterized further.

gene, and spontaneous gene amplification increases gene dosage, with consequent increases in gene expression. In pursuit of the last of these afflictions, we have uncovered what may be a previously unrecognized protooncogene whose amplification may contribute to the genesis of human neuroblastomas. We have called this gene *N-myc* because it is a distant kin of the protooncogene *c-myc* (the kinship led to the discovery of *N-myc*) (*12*).

N-myc is amplified in ~50% of human neuroblastomas classified as stage III or IV (*13*), a classification that denotes advanced and incurable disease. The amplification is a somatic lesion, since it is found only in tumors, not in adjacent normal tissue (*14*). Enhanced expression accompanies gene amplification (*14*) but has not been found in its absence (M. Schwab, unpublished observations).

We suggest that amplification of *N-myc* represents one of the factors that can lead to malignant progression of the tumor cells destined to engender the advanced stages of neuroblastoma. The suggestion would have greater credence, however, if there were evidence that *N-myc* has biological activity typical of protooncogenes. We therefore turned to the assay invented by Weinberg, Ruly, and colleagues, in which two oncogenes are used cooperatively to transform primary explants of rat embryo fibroblasts (*6, 15*). A transcriptional promoter–enhancer from a murine retrovirus was used to guarantee brisk expression of *N-myc* during transfection. A combination of mutant *c-ras* (derived from the EJ line of human bladder carcinoma cells) with *N-myc* transformed rat cells to tumorigenic growth (Table IV) (M. Schwab, unpublished observations). These findings indicate that the physiological effects of N-*myc* are akin to those of *c-myc*, large T antigen of polyoma virus, and the *E1a* gene of adenovirus, all of which also cooperate with mutant *c-ras* as described (*6, 15*). Emboldened by these findings, we can now suggest that *N-myc* qualifies as a protooncogene, and that its role in the genesis of neuroblastoma deserves diligent exploration.

V. Conclusion

The work reviewed above exemplifies how the study of retroviral oncogenes and their cellular parents has wedded the investigation of carcinogenesis to the study of cellular and developmental biology. The thought that differentiation and cancer are intertwined is not new. What is new is the identification of genes whose activities may be germane to both the control of differentiation and the genesis of neoplastic growth. Our findings with *N-myc* are illustrative of this

TABLE IV

***N-myc* as an Oncogene in Primary Rat Embryo Cultures[a]**

DNA transfected	Indefinite growth	Morphological transformation	Anchorage independence	Tumorigenicity in *Nu/Nu*
N-myc	−	−	−	−
N-myc + *EJ-ras*	−	−	−	−
LTR/*N-myc*	−	−	−	−
LTR/*N-myc* + *EJ-ras*	+	+	+	+

[a] Molecular clones whose composition is summarized in the table were introduced into primary cultures of rat embryo cells by the use of calcium phosphate precipitates. Cultures that showed no consequences of transfection were scored as (−). When foci of morphological transformed cells were observed (+), these were propagated for evaluation of the other phenotypic properties given in the table. LTR denotes the long terminal repeat of Moloney murine leukemia virus, which contains a transcriptional promoter and enhancer. It was joined to a molecular clone of *N-myc* at a naturally occurring restriction site. *EF-ras* is a mutant version of *c-ras* isolated from the EJ line of human bladder carcinoma cells.

conjunction: in its normal guise, *N-myc* may be a gene whose activity is restricted to cells of the neuroectodermal lineage (*14*); amplification and exaggerated expression of *N-myc*, however, may urge the cells of human neuroblastoma toward an ever more threatening phenotype (*13*).

Where do we go from here? We must first confront the fact that genetic lesions have so far gone undetected in the large majority of human tumors. We may need to invent new means by which to search for these lesions, and we must remain open to the possibility that we will not always find them because they are not always there. We also need to enlarge our purview to accomodate the fact that some tumors are afflicted by genetic damage whose manifestation is recessive. There is at present no evidence for how such damage might interact with oncogenes.

The search for yet more oncogenes will continue. We also hope to learn how widely our genetic paradigm can be applied. Through the study of oncogenes, can we find explanations for inherited diatheses to cancer? Can we identify initial targets for carcinogens, can we spot the lesions that initiate tumorigenesis? Are there genetic explanations for the phenotypic progression of tumor cells towards greater malignancy? And by what means—oncogenes or others—is the final malignant growth of cancer cells sustained? This last issue raises the most vital issue of all. We remain largely ignorant of how the proteins encoded by oncogenes act. We must banish that ignorance before we can claim to know the inner workings of the cancer cell, before we can hope to parlay the explication of oncogenes into decisive strategies for the treatment and prevention of cancer.

Acknowledgments

Supported by the National Institutes of Health, the American Cancer Society, and private donors to the G. W. Hooper Research Foundation. The author thanks his colleagues for the privilege of citing unpublished observations, and Lynn Vogel for preparing the manuscript.

References

1. J. M. Bishop, *Ann. Rev. Biochem.* **52,** 301 (1983).
2. T. Hunter, *Sci. Am.* **251(2)**, 70 (1984).
3. E. M. Durban and D. Boettiger, *Proc. Natl. Acad. Sci. U.S.A.* **78,** 3600 (1981).
4. R. Ralston and J. M. Bishop, *Nature (London)* **306,** 803 (1983).
5. P. Donner, T. Bunte, I. Greiserwilke, and K. Moelling, *Proc. Natl. Acad. Sci. U.S.A.* **80,** 2861 (1983).
6. H. Land, L. F. Parada, and R. A. Weinberg, *Nature (London)* **304,** 596 (1983).
7. G. Symonds, K.-H. Klempnauer, G. Evan, and J. M. Bishop, *Mol. Cell. Biol.* in press (1984).
8. K.-H. Klempnauer, G. Symonds, G. I. Evan, and J. M. Bishop, *Cell* **37,** 537 (1984).
9. B. Z. Shilo and R. A. Weinberg, *Proc. Natl. Acad. Sci. U.S.A.* **78,** 6789 (1981).
10. M. A. Simon, T. B. Kornberg, and J. M. Bishop, *Nature (London)* **302,** 837 (1983).
11. H. E. Varmus, *Ann. Rev. Genet.* **18,** 553 (1984).
12. M. Schwab, K. Alitalo, K.-H. Klempnauer, H. E. Varmus, J. M. Bishop, F. Gilbert, G. Brodeur, M. Goldstein, and J. Trent, *Nature (London)* **305,** 245 (1983).
13. G. M. Brodeur, R. C. Seeger, M. Schwab, H. E. Varmus, and J. M. Bishop, *Science* **224,** 1121 (1984).
14. M. Schwab, J. Ellison, M. Bush, W. Rosenau, H. E. Varmus, and J. M. Bishop, *Proc. Natl. Acad. Sci. U.S.A.* **81,** 4940 (1984).
15. H. E. Ruly, *Nature (London)* **304,** 602 (1983).

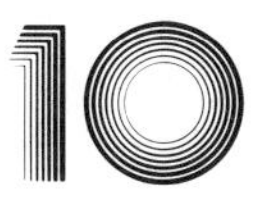

Differences in the Biological Function of Viral and Cellular *src* Genes

HIDEO IBA, TATSUO TAKEYA, FREDERICK R. CROSS, ELLEN A. GARBER, TERUKO HANAFUSA, DAVID PELLMAN, AND HIDESABURO HANAFUSA

The Rockefeller University
New York, New York

I. Introduction

The incorporation of cellular protooncogene sequences into retroviral genomes is one of the mechanisms known to create active oncogenes. Comparative analysis of some viral oncogenes and their cellular counterparts has shown that inactive protooncogenes convert to active oncogenes in various ways. In some cases, the mere overexpression of protooncogenes under the control of the strong viral promoter seems to be sufficient for the conversion to oncogenes (*1–3*). In other systems, mutation rather than transcriptional activation has been found to be critical for the conversion (*4–6*). In one case, only a part of the protooncogene was incorporated into the viral genome to become an active oncogene (*7*). In this case, the incorporated portion of the protooncogene may be sufficient for transforming activity, but it is still possible that the removal of a portion of the protooncogene may endow the transduced gene with an altered function. Therefore, comparison of a viral oncogene and its cellular counterpart would contribute to understanding of the conditions for the activation of a protooncogene and also may reveal the functionally important molecular structure of the oncogene product. In this report, we will describe our findings which indicate that the cellular *src* (*c-src*) gene requires mutations in the coding sequence to become the transforming viral *src* (*v-src*) gene (*8, 9*) and that the overproduced *c-src* gene product is unable to cause cell transformation becuase this protein has an extremely low protein kinase activity (*10*).

II. Lack of Transforming Activity of the Overexpressed *c-src* Gene

Previous analysis of the *v-src* gene of Rous sarcoma virus (RSV) and its cellular homolog, the chicken *c-src* gene, demonstrated typical structural differences between a viral and a eukaryotic gene, the former being present as a processed messenger RNA structure (*11–13*). The *c-src* locus contains 11 introns, and only 1.6 kb of the 7 kb gene is the coding region. However, the coding region of *c-src* and *v-src* share a remarkable similarity in both nucleotide and amino acid sequences. Of 533 amino acids of the *c-src* protein, $p60^{c\text{-}src}$, only seven single amino acid substitutions are found in the sequence compared to the *v-src* protein, $p60^{v\text{-}src}$, except for their carboxyl termini (Fig. 1) (*13*). We found that the last 19 amino acids of the carboxyl end of $p60^{c\text{-}src}$ are replaced by 12 different amino acids in $p60^{v\text{-}src}$. The sequence encoding these 12 amino acids of $p60^{v\text{-}src}$ is present about 1 kb downstream from the termination codon of the *c-src* gene, and this sequence must have substituted for the 3′ end of the *c-src* sequence when the *v-src* sequence was formed (*13*).

The expression of the *c-src* gene was previously shown to be extremely limited compared with that of the *v-src* gene (*14, 15*), and a hypothesis was proposed that the difference in the transcriptional levels may explain why normal cells containing *c-src* are not transformed (*16*). The restoration of transforming activity by transformation-defective mutants that have large deletions in the *src* gene after recombination with the *c-src* sequence seems to support this hypothesis and also suggested a functional similarity between the products of *v-src* and

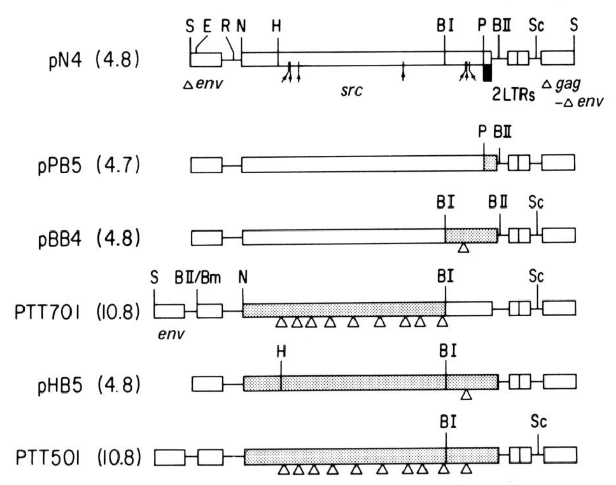

Fig. 1. The structure of p60 proteins encoded by various plasmids constructed. In the diagram of the *src* gene, white boxes and shaded boxes indicate the *v-src* sequence and the *c-src* sequence, respectively. Triangles (△) show the sites of the introns present in the nucleotide sequence of *c-src*. The arrows under the *v-src* of N4 show the positions of 7 amino acid differences between *v-src* and *c-src,* and the black box shows the locations of the C-terminal 12 amino acids of *v-src,* which are different from the C-terminal 19 amino acids of *c-src*. LTR, Long terminal repeat, a transcriptional promoter. The abbreviations for restriction enzymes are as follows: *Sal*I (S), *Eco*RI (E), *Rsa*I (R), *Nco*I (No), *Hga*I (H), *Bgl*I (BI), *Pst*I (P), *Sac*I (Sc), and *Bam*HI (Bm).

c-src (*17, 18*). However, this rescue of the transforming function by recombination with the endogenous *c-src* sequence was possible only with partially deleted mutants (the deletion of the whole *v-src* sequence prevents the formation of recombinant sarcoma virus) (*19*). Furthermore the recovery of the recombinant virus was dependent on the detection of tumors that could be induced by mutants formed subsequent to the recombination. Therefore, these experiments do not allow us to conclude whether virus containing only the *c-src* sequence is active in cell transformation.

In order to determine whether the overexpression of the *c-src* gene can cause cell transformation, we constructed RSV variant DNAs *in vitro* in which the *v-src* sequence was partially or entirely replaced with the *c-src* sequence (*8, 9*). The structures of these constructed DNAs are shown in Fig. 1. These DNAs contain part of the *env* gene, the entire *src* gene, and 3′-LTR (long terminal repeat). Since chicken cells are not transformed with retrovirus DNA unless subsequent virus multiplication takes place, transfection of chicken embryo fibroblasts (CEF) with these constructed DNAs alone does not cause transformation even with the plasmid pN4, which contains the wild-type *v-src* gene. They were therefore ligated, before transfection, to the plasmid pSR-REP, which contained the remaining RSV sequence required for viral replication (*20*). The resulting DNA is colinear with the nondefective form of RSV proviral DNA. Since transfection with these ligated DNAs produces infectious virus regardless of their cell-transforming activity, the success of transfection can be monitored by the assay of reverse transcriptase activity, which reflects the virus released into the culture. Chicken cell cultures became transformed within 7–9 days following transfection with pN4. Both the variants pBB4, which should encode a p60 consisting of the N-terminal amino acids 1–431 of $p60^{v\text{-}src}$ and the C-terminal amino acids 432–533 of $p60^{c\text{-}src}$, and another variant, pTT701, which encodes a p60 with the chimeric structure reciprocal to pBB4 p60, induced cell transformation with essentially the same time course as did the wild-type pN4. pPB5, which encodes a p60 consisting of amino acids 1–513 of $p60^{v\text{-}src}$ and 514–533 of $p60^{c\text{-}src}$, was also active in transformation. However, pTT501 and pHB5, which encode the entire $p60^{c\text{-}src}$, were unable to transform CEF, although they did produce high titers of nontransforming virus.

All transfected cultures produced an equal level of reverse transcriptase by 7 days, and the cultures contained the expected genomic size of DNA, in which all introns present in the transfecting DNA were removed. The analysis of the protein product of the *src* gene by immunoprecipitation of cell lysates with antisera against $p60^{src}$ showed that all cultures, including those transfected with DNAs that contained the entire *c-src* sequence, produced the same level of $p60^{src}$. From these results, we concluded that the transfection was equally efficient in all cultures, resulting in the production of virus that was spread and induced the formation of $p60^{src}$. The viruses obtained were used for the analysis of the properties of $p60^{src}$. These were designated as NYBB4 when recovered from cultures transfected with pBB4, etc.

While the cultures transfected with plasmids containing only the *c-src* sequence are not transformed, a few foci were observed in successive subcultures. Since these foci were producing highly transforming virus, we concluded that the formation of foci is due to mutant viruses arising from the nontransforming *c-src*–containing viruses. The frequency of the mutation appears to be quite high (10^{-3}–10^{-4} of total infectious virus).

Thus the results indicate that the *c-src* sequence is inactive in causing cell transformation even when it is overexpressed but once this sequence is incorporated in a retrovirus, mutant transforming virus is formed at a relatively high rate. The fact that the two reciprocal chimeric viruses, NYBB4 and NY701, can transform cells suggests that the *src* gene of the RSV strain used contains at least two mutations in relation to the *c-src* gene, one upstream and one downstream of the *Bgl*I site, and that either one alone is sufficient to cause transformation. Whether these mutations represent hot spots or they are one of many possible sites whose mutations affect the transforming activity of the product has not been determined.

III. The Basis of the Inability of p60$^{c\text{-}src}$ to Transform Cells

The viral *src* gene product p60$^{v\text{-}src}$ is known to have a tyrosine-specific protein kinase activity, and this activity appears to be correlated to the transforming activity of this protein (*21, 22*). It is also known that p60$^{v\text{-}src}$ is associated with the plasma membrane. From the analysis of various viral mutations in which mutations were introduced in the N-terminal 9 kdal of p60$^{v\text{-}src}$, we have recently obtained evidence that fatty acid (myristic acid) attachment to the N-terminal region of p60$^{v\text{-}src}$ is required for its association with the plasma membrane and for its activity to transform cells (*23*). As shown in Table I, mutants in which the N-terminal sequences from amino acids 15–27, 15–49, and 15–81 were deleted (and substituted with unrelated di- to tetrapeptides) were active in transformation but all mutants in which amino acid 2 (glycine residue) is modified were inactive in transformation. In every constructed mutant, the ability of cell transformation was correlated to the capacity to bind myristic acid at N-terminal glycine after the removal of methionine 1 and to associate with the plasma membrane in infected cells (*23, 24*).

The analysis of myristic acid binding of p60$^{c\text{-}src}$, which was produced in NY501-infected cells, showed that the overproduced p60$^{c\text{-}src}$ bound myristic acid and associated with the membrane, as did wild-type p60$^{v\text{-}src}$. Therefore, the inactivity of the overproduced p60$^{c\text{-}src}$ in transformation cannot be attributed either to properties related to the structure of the N terminus or to the subcellular localization.

The protein kinase activity of p60^{src} is generally measured by an *in vitro* assay using an immune complex formed by p60, the antibody, and protein A–Sepharose (*16, 25*). If, however, the conformation of the antigen is altered by the

TABLE I

Characterization of RSV Variants Whose p60 Proteins Have Mutations in the N-Terminal Region

Virus	Amino acids deleted	Amino acids substituted	Transformation	Kinase[a] (IgG)	Membrane association	Myristylation
SRA			+	1.0	+	+
NY307	15–27	PQIW	+	1.0	+	+
NY308	15–49	PRSG	+	1.1	+	+
NY309	15–81	PDL	+	0.8	+	+
NY314	2–81	DL	–	0.6	–	–
NY315	2–15	DLG	–	1.0	–	–

[a] *src* protein kinase assay measured by TBR IgG phosphorylation (fraction of SRA value).

binding with antibody, the enzyme activity of the antigen would be affected. To avoid these possible artifacts of the reaction in the immune complex, phosphorylation of proteins *in vivo* (in infected cells) was used for the evaluation of the kinase activity of the overproduced $p60^{c\text{-}src}$. We found that infection with NY501 or NYHB5 only slightly elevated the levels of both phosphorylation of the total cell protein at tyrosine residues (*26*) and that of cellular 34,000 protein (*27, 28*) over the levels in uninfected cells, whereas these are increased in wild-type NYN4 virus–infected cells up to nine and 3.5 times, respectively (Table II). (The phosphorylation of 34,000 protein in uninfected or *c-src*–containing virus-infected cells is essentially on serine residues.) p60 proteins produced by the RSV variants such as NYPB5 and NYBB4, which have a chimeric structure containing N-terminal *v-src* and C-terminal *c-src*, were highly active in kinase activity, while the NY701 p60, containing N-terminal *c-src* and C-terminal *v-src*, had 50–70% of the specific activity of the wild-type $p60^{v\text{-}src}$. Interestingly, p60 proteins of the transforming viruses generated by spontaneous mutations of the *c-src*–containing viruses (NYHB5-T7, NYHB5-T9, NY501-T7, and NY501-T9) have specific activities in a range similar to that of NY701 p60.

We also analyzed the site of tyrosine phosphorylation of these p60 proteins ^{32}P-labeled *in vivo*. Previous studies have shown that the major tyrosine phos-

TABLE II

Protein Kinase Activity of Variant RSV p60 Proteins *in Vivo*

Virus	Transformation	P-Tyr in total cell protein (%)	32P in 34,000 protein (cpm)	Phosphorylation of p60 at Tyr(416)
NYN4	+	0.67	560	+
NYPB5	+	0.87	860	+
NYBB4	+	0.83	1040	+
NY701	+	0.48	260	+
NYHB5-T8	+	0.27	240	+
NYHB5-T9	+	0.43	360	+
NY501-T7	+	0.42	580	+
NY501-T9	+	0.25	310	+
NYHB5	–	0.13	140	Trace
NY501	–	0.12	170	Trace
Uninfected	–	0.072	160	Trace

phorylation occurs at the tyrosine residue at amino acid position 416 of $p60^{v\text{-}src}$, but at another undetermined position of $p60^{c\text{-}src}$ in uninfected cells (*29, 30*). We found that $p60^{c\text{-}src}$ in NY501- or NYHB5-infected cells is phosphorylated at the same site as endogenous $p60^{c\text{-}src}$ in uninfected cells. Tyrosine 416, however, was a major site of phosphorylation in $p60^{src}$ proteins formed by spontaneous transforming mutants derived from the *c-src*–containing viruses.

These results indicate that even when it is overproduced, $p60^{c\text{-}src}$ remains essentially inactive in tyrosine kinase activity compared with $p60^{v\text{-}src}$ and cannot induce tyrosine phosphorylation of cellular proteins, as well as tyrosine 416 of itself. Thus, the low level of kinase activity of the overproduced $p60^{c\text{-}src}$ appears to be primarily responsible for the inability of the *c-src*–containing virus to transform. In spontaneous mutants, acquired transforming activity was always accompanied by an increased specific activity in the protein kinase. The difference in the site of tyrosine phosphorylation in the product of the *v-src* and *c-src* gene may reflect a difference in their conformation.

IV. Function of Endogenous $p60^{c\text{-}src}$

The biological function of the *c-src* gene product is still unknown. Since the protein kinase activity of overproduced $p60^{c\text{-}src}$ is very low in CEF, as described above, it is highly possible that its activity is tightly regulated by other cellular proteins or cofactors. In the case of other cellular tyrosine kinases such as the epidermal growth factor (EGF) receptor, the insulin receptor, and the platelet-derived growth factor (PDGF) receptor, the activity is regulated by the binding of the appropriate protein growth factors (*31*). $p60^{v\text{-}src}$ may be analogous to the gene product of *v-erbB,* which is derived from the gene for EGF receptor (*7*), in being an unregulated version of a normally regulated cellular tyrosine kinase. Although polyoma middle T antigen is not a cellular protein, it is interesting that this viral protein can tightly bind to $p60^{c\text{-}src}$ (*32*) and activate the kinase activity of $p60^{c\text{-}src}$ (*33*). The understanding of such systems, which regulate the kinase activity of $p60^{c\text{-}src}$, may shed light on the function of endogenous $p60^{c\text{-}src}$.

Acknowledgments

This work was supported by Public Health Service Grants CA 14935 and CA 18213 from the National Cancer Institute and by Grant MV128A from the American Cancer Society. F. R. C. was supported by NIH Training Grant AI07233, and E. A. G. was a recipient of a Merck fellowship.

References

1. M. Oskarsson, W. L. McClements, D. G. Blair, J. V. Maizel, and G. F. Vande Woude, *Science* **212,** 941 (1981).
2. E. H. Chang, M. E. Furth, E. M. Scolnick, and D. R. Lowy, *Nature (London)* **297,** 479 (1982).
3. A. D. Miller, T. Curran, and I. M. Verma, *Cell* **36,** 51 (1984).
4. L. F. Parada, C. J. Tabin, C. Shih, and R. A. Weinberg, *Nature (London)* **297,** 474 (1982).
5. E. Santos, S. R. Tronick, S. A. Aaronson, S. Pulciani, and M. Barbacid. *Nature (London)* **298,** 343 (1982).
6. C. Der, T. Kronitiris, and G. Cooper, *Proc. Natl. Acad. Sci. U.S.A.* **79,** 3637 (1982).
7. J. Downward, Y. Yorden, E. Mayes, G. Scrace, N. Totty, P. Stockwell, A. S. Ullich, J. Schlessenger, and M. D. Waterfield, *Nature (London)* **307,** 521 (1984).
8. H. Hanafusa, H. Iba, T. Takeya, and F. R. Cross, *in* "Cancer Cells" (G. F. Vande Woude, A. J. Levine, W. C. Topp, and J. D. Watson, eds.), Vol. 2, P. 1. Cold Spring Harbor Laboratory, New York, 1984.
9. H. Iba, T. Takeya, F. R. Cross, T. Hanafusa, and H. Hanafusa, *Proc. Natl. Acad. Sci. U.S.A.* **81,** 4424 (1984).
10. H. Iba, F. R. Cross, E. A. Garber, and H. Hanafusa, *Mol. Cell. Biol.* **5,** 1058 (1985).
11. D. Shalloway, A. D. Zelenety, and G. M. Cooper, *Cell* **24,** 531 (1981).
12. R. C. Parker, H. E. Vermus, and J. M. Bishop, *Proc. Natl. Acad. Sci. U.S.A.* **78,** 5842 (1981).
13. T. Takeya and H. Hanafusa, *Cell* **32,** 881 (1983).
14. D. H. Spector, K. Smith, T. Padgett, P. McCombe, D. Roulland-Dussoix, C. Moscovici, H. E. Vermus, and J. M. Bishop, *Cell* **13,** 371 (1978).
15. S. Y. Wang, W. S. Hayward, and H. Hanafusa, *J. Virol.* **24,** 64 (1977).
16. R. E. Karess, W. S. Hayward, and H. Hanafusa, *Proc. Natl. Acad. Sci. U.S.A.* **76,** 3154 (1979).
17. H. Hanafusa, C. C. Halrern, D. L. Buchhagen, and S. Kawai, *J. Exp. Med.* **146,** 1735 (1977).
18. L.-H. Wang, C. C. Halpern, M. Nadel, and H. Hanafusa, *Proc. Natl. Acad. Sci. U.S.A.* **75,** 5812 (1978).
19. L.-H. Wang, M. Beckson, S. M. Anderson, and H. Hanafusa, *J. Virol.* **49,** 881 (1984).
20. F. R. Cross and H. Hanafusa, *Cell* **34,** 597 (1983).
21. B. M. Sefton, T. Hunter, K. Beemon, and W. Eckhart, *Cell* **20,** 807 (1980).
22. D. L. Bryant and T. J. Parsons, *Mol. Cell. Biol.* **4,** 862 (1984).
23. F. R. Cross, E. A. Garber, D. Pellman, and H. Hanafusa, *Mol. Cell. Biol.* **4,** 1834 (1984).
24. D. Pellman, E. A. Garber, F. R. Cross, and H. Hanafusa, *Proc. Natl. Acad. Sci. U.S.A.* **82,** 1623 (1985).
25. J. S. Brugge, E. Erikson, and R. L. Erikson, *Cell,* **25,** 363 (1981).
26. T. Hunter and B. W. Sefton, *Proc. Natl. Acad. Sci. U.S.A.* **77,** 1311 (1980).
27. K. Radke and G. S. Martin, *Proc. Natl. Acad. Sci. U.S.A.* **76,** 5212 (1979).
28. E. Erikson and R. L. Erikson, *Cell* **21,** 829.
29. J. E. Smart, H. Oppermann, A. P. Czernilofsky, A. F. Purchio, R. L. Erikson, and J. M. Bishop, *Proc. Natl. Acad. Sci. U.S.A.* **78,** 6013 (1981).
30. R. E. Karess and H. Hanafusa, *Cell* **24,** 1550 (1981).
31. C.-H. Heldin and B. Westermark, *Cell* **37,** 9 (1984).
32. S. A. Courtneidge and A. E. Smith, *Nature (London)* **303,** 435 (1983).
33. J. B. Bolen, C. J. Thiele, M. A. Israel, W. Yonemoto, L. A. Lipsich, and J. S. Brugge, *Cell* **38,** 767 (1984).

11

Function of Yeast *RAS* Genes

MICHAEL WIGLER, SCOTT POWERS, TOHRU KATAOKA, TAKASHI TODA, OTTAVIO FASANO

Cold Spring Harbor Laboratory
Cold Spring Harbor, New York

KUNIHIRO MATSUMOTO

Department of Industrial Chemistry
Tottori University
Tottori, Japan

ISAO UNO, TATSUO ISHIKAWA

Institute of Applied Microbiology
University of Tokyo
Tokyo, Japan

JAMES BROACH

Department of Molecular Biology
Princeton University
Princeton, New Jersey

I. Introduction

The *ras* genes were first discovered as the oncogenes of the Harvey and Kirsten rat sarcoma viruses (*5*). Highly conserved in evolution, this gene family encodes guanine nucleotide binding proteins (*14, 20, 21*), which become associated with the plasma membrane (*22*). "Activated" *ras* genes, capable of the tumorigenic transformation of NIH3T3 cells, are found in many human and rodent tumor cells and differ from the normal *ras* genes by single missense mutations (*1, 17, 23, 25, 27, 28, 33*). The normal *ras* proteins have a weak GTPase activity, and the level of this activity in oncogenic variants is greatly reduced (*13, 24*). The cellular function of the normal *ras* proteins in vertebrates is unknown.

In the yeast *Saccharomyces cerevisiae* there are two genes, *RAS1* and *RAS2*, which are closely homologous to the mammalian *ras* genes (*3, 4, 16*). We have been studying the function of the yeast *RAS* gene in search of clues about mammalian *ras* function. In the course of these studies, we discovered that mammalian *ras* protein can function in yeast cells (*30*) and that the yeast *RAS* genes are essential controlling elements of adenylate cyclase (*9*).

II. Yeast *RAS* Genes Are Essential for Growth and Viability

We and others have previously shown that at least one functional *RAS* gene is essential for the germination of haploid yeast spores (*8, 29*). This was demonstrated by constructing doubly heterozygous diploid yeast cells containing wild-type *RAS1* and *RAS2* alleles and *RAS1* and *RAS2* alleles each disrupted by a different auxotrophic marker. These diploid cells were then sporulated and the resulting tetrads were analyzed. Only spores containing at least one functional *RAS* gene could germinate. We then introduced into these doubly heterozygous diploid cells a *RAS2* gene under the transcriptional control of the galactose-inducible *GAL10* promoter (*30*) linked to a third auxotrophic marker. These diploid cells were then sporulated and tetrads germinated on either glucose-containing medium (YPD) or galactose-containing medium (YPGal) (Table I). The results confirm that at least one functional *RAS* gene is required for germination. *ras1*$^-$ *ras2*$^-$ *GAL10–RAS2* spores germinate only in the presence of galactose, the inducer for the *GAL10* promoter.

TABLE I

Spore Viability in Tetrads from JR33[a]

Spore genotype			Germination			
			on YPD		on YPGal	
RAS1	*RAS2*	*GAL10–RAS2*	Viable	Nonviable	Viable	Nonviable
+	+	+	7	0	10	0
+	+	−	9	1	13	1
+	−	+	9	0	10	1
+	−	−	6	2	9	0
−	+	+	9	0	9	2
−	+	−	7	0	10	0
−	−	+	0	6	13	1
−	−	−	0	7	0	10

[a] A detailed description of the genotype of JR33 is given in Ref. 9. The genotype for viable spores was determined by the presence of auxotrophic markers; for nonviable spores, the genotype was determined from the genotype of viable spores within a tetrad, assuming Mendelian genetics.

Cells with the genotype *ras1*⁻ *ras2*⁻ *GAL10–RAS2* were grown in YPGal and then shited to YPD. The growth of these cells was then monitored and cultures were plated onto YPGal agar to measure cell viability. Within several cell generations, cell growth ceased and cells lost viability (data not shown). These experiments indicated that not only are the *RAS* genes needed for spore germination, but they are also needed for the continued growth and viability of yeast cells.

III. Mammalian *H-ras* Complements *ras1*⁻ *ras2*⁻ Yeast

Using the approach described above, we have been able to test if expression of the normal mammalian *H-ras* protein is sufficient for viability in yeast cells lacking their own endogenous *RAS* genes. To this end we constructed a *GAL10–H-ras* transcription unit that utilized a full length cDNA clone of the human *H-ras* mRNA under the control of the galactose-inducible *GAL10* promoter. This unit, closely linked to a *LEU2* marker, was inserted into diploid yeast cells, which were doubly heterozyous for their endogenous *RAS* genes. Cells were

TABLE II

Spore Viability in Tetrads from JR34[a]

Spore genotype			Germination			
			on YPD		on YPGal	
RAS1	*RAS2*	*GAL10–H-ras*^{gly12}	Viable	Nonviable	Viable	Nonviable
+	+	+	6	0	15	0
+	+	−	7	0	29	0
+	−	+	6	0	20	0
+	−	−	4	0	18	1
−	+	+	3	0	22	0
−	+	−	6	2	17	2
−	−	+	0	10	10	16
−	−	−	0	4	0	16

[a] The detailed description of the genotype of JR34 is given in Ref. *9*. The genotypes of spores were determined as described in Table I.

induced to sporulate and tetrads examined after germination on YPD or YPGal plates (Table II). Approximately 40% of spores with the genotype *ras1*$^-$ *ras2*$^-$ *GAL10–H-ras* were capable of germination when plated on YPGal, from which we conclude that the human *H-ras* protein can supply essential *RAS* function to yeast.

The relatively poor germination of *ras1*$^-$ *ras2*$^-$ *GAL10–H-ras* spores indicates that, although the *H-ras* protein functions in yeast, it does so inefficiently. Moreover, the colonies resulting from such spores were clearly smaller than wild type colonies, suggesting growth was impaired. This was confirmed by performing growth curves. Microscopic examination of cells in exponential growth showed that cells with only the human *H-ras* protein were predominantly unbudded and therefore had a prolonged G_1 phase (Table III). *RAS* function is evidently required for efficiently traversing the G_1 phase of the cell cycle.

IV. Yeast Cells Carrying *RAS2*^{val19} Have a Defective Response to Nutritional Stress

Certain missense mutations drastically alter the biological activity of mammalian *ras* genes. In particular the *H-ras*^{val12} gene, which encodes valine instead of glycine at the twelfth codon of the human *H-ras* gene, can induce the tu-

TABLE III

Doubling Time and Duration of G_1 in Various Strains[a]

Strain	Genotype			Doubling time, D (hr)	Fraction unbudded, F	G_1 (hr)	$D-G_1$ (hr)
	RAS1	*RAS2*	*GAL10–H-ras*				
KPPK-3A	–	+	–	2.5	.48	0.99	1.51
JR34-14A	–	–	+	4.8	.73	3.14	1.66
JR34-11C	–	–	+	5.6	.74	3.73	1.87
JR34-3B	–	–	+	6.0	.75	4.07	1.93

[a] The full genotypes of the indicated strains are given in Ref. *9*. The doubling time D of cells inoculated in rich medium containing galactose (YPGal) was measured during exponential growth. The fraction F of unbudded cells (cells in G_1) during exponential growth was determined by the microscopic examination of at least 100 cells. The duration of G_1 was determined using the formula $G_1 = D(1 - \log(2 - F)/\log 2)$ (*18*). The difference between D and G_1 ($D - G_1$) represents the duration of the cell cycle, excluding G_1.

morigenic transformation of NIH3T3 cells. To test the consequences of a similar mutation of *RAS2* on the properties of yeast cells, we constructed the *RAS2*^{val19} gene using site-directed mutagenesis (*8*). *RAS2*^{val19} thus encodes valine instead of glycine at position 19, which corresponds to position 12 of the mammalian *H-ras* protein. We then examined the effects of introducing this gene into yeast cells.

Our first observation was that diploid cells containing *RAS2*^{val19} could not be induced to sporulate by incubation under conditions of nutritional deprivation (*8*). Next we observed that haploid cells carrying *RAS2*^{val19} lost viability if starved for nitrogen, sulfur, or phosphorus and failed to arrest in G_1 under those conditions (Table IV). Wild-type cells normally arrest in G_1 and retain viability for long periods if nutritionally deprived. In addition to these phenotypes, we also noted that cells carrying *RAS2*^{val19} failed to accumulate carbohydrate stores as cells entered the stationary growth phase (*30*). Thus, in general, *RAS2*^{val19} cells appeared to show a defective response to nutritional stress.

V. *bcy1* Is a Suppressor Mutation for *RAS* Disruptions

The cluster of cellular properties just described closely resemble the phenotype of cells carrying the *bcy1* mutation (*10–12, 31*). The *bcy1* mutation was first

TABLE IV

Viability of Strains under Various Conditions

	Cell Strain[a]			
	TK161-R2G		TK161-R2V	
Growth condition[b]	Budded[c] (%)	Viable[c] (%)	Budded[c] (%)	Viable[c] (%)
YPD	12	100	56	80
SD complete	25	100	32	5.
−nitrogen	0	100	60	0.1
−auxotrophic requirements	11	13	40	0.01
	TK161-R1G		TK161-R1V	
YPD	18	100	35	70
SD complete	6	90	35	5.
−nitrogen	2	40	30	1.
−sulfate	1	90	14	2.

[a] The four strains listed here are described fully in Ref. *30*. Strain TK161-R2V is isogenic to TK161-R2G but contains the *RAS2*^{val19} allele. Similarly, Tk162-R2V is isogenic to TK162-R2G but contains the *RAS2*^{val19} allele.

[b] TK161-R2G and TK161-R2V were inoculated from log phase cultures into the indicated liquid medium at 3×10^5 cells/ml and incubated 4 days. TK162-R1G and TK162-R1V were inoculated at 10^6 cells/ml and incubated for 36 hr. Culture conditions were either rich medium (YPD) or synthetic medium (SD) supplemented with the required auxotrophic supplements (complete) or without (−auxotrophic requirements). For nitrogen starvations, ammonium sulfate and all auxotrophic requirements were omitted. For sulfate starvation, ammonium and magnesium sulfates were replaced with ammonium and magnesium chlorides. All cultures had reached stationary phase in YPD and in complete SD.

[c] The percentage of budded cells was determined by the microscopic examination of at least 200 cells. The percentage of viable cells was determined by the colony-forming efficiency of sonicated cultures on YPD agar.

isolated by Matsumoto and co-workers as a mutation that suppresses the lethality that otherwise results from the disruption of adenylate cyclase (*10*). Cells carrying *bcy1* appear to lack the regulatory subunit of the cAMP-dependent protein kinase and hence have lost the requirement for cAMP (*32*). These observations suggested that the *RAS* genes might be participating in the cAMP pathway. In support of this idea we found that *bcy1* suppressed the lethality that otherwise results from disruption of both *RAS* genes (Table V).

TABLE V

Tetrad Dissection of *ras1*/+ *ras2*/+ *bcy1*−/+ Diploids[a]

Cross			T16-2A/S5S2		T16-3D/S5S2		T17-7B/T3-23B	
RAS1	*RAS2*	*BCY1*	Viable	Nonviable	Viable	Nonviable	Viable	Nonviable
+	+	+	10	0	5	0	14	0
−	+	+	13	0	14	1	16	1
+	−	+	20	0	14	0	11	1
−	−	+	0	15	0	4	0	4
+	+	−	16	0	7	0	6	2
−	+	−	17	0	10	1	9	1
+	−	−	11	0	9	1	13	0
−	−	−	4	7	5	4	8	10

[a] See Ref. *9* for complete description of strains. Individual diploids from the indicated crosses were sporulated and tetrads dissected. The genotypes of all viable spores were determined as follows: the *RAS* phenotypes were deduced from the presence of auxotrophic markers used to disrupt the respective genes. The *bcy1* phenotypes were deduced from the cluster of phenotypes that identify this mutation and segregate in 2:2 fashion. When possible, the genotypes of nonviable spores were assigned on the basis of the viable spores within a tetrad, assuming normal Mendelian segregation of genetic loci. The table summarizes data from tetrads in which complete genotypic determinations only were possible. Overall spore viability was 65%.

Yeast *RAS* Genes Modulate Adenylate Cyclase

The adenylate cyclase activity of the yeast *S. cerevisiae* is stimulated by guanine nucleotides in the presence of magnesium (*2*). In this respect, yeast adenylate cyclase resembles the adenylate cyclase of mammalian cells, which can be stimulated by a guanine nucleotide–binding complex called G_S (*6*). Since the yeast *RAS* proteins also bind guanine nucleotides (*26*), they might also modulate adenylate cyclase. This was tested directly by the assay of membranes from wild-type yeast cells and *ras1*$^-$ *ras2*$^-$ cells (Table VI). Membranes from either contained appreciable adenylate cyclase activity when assayed in the presence of manganese, but *ras1*$^-$ *ras2*$^-$ membranes displayed negligible levels of activity when assayed in the presence of magnesium and a nonhydrolyzable guanine nucleotide analog.

These rsults were confirmed in a striking manner by membrane-mixing experi-

TABLE VI

Adenylate Cyclase Activity in Membranes

Strain	Genotype	Assay conditions		
		Mn^{2+}	Mg^{2+}	Mg^{2+}, Gpp(NH)p
Experiment 1[a]				
SP1	*RAS1 RAS2*	49.2	3.7	14.3
TK161-R2V	*RAS1 RAS2*^{val19}	51.9	14.2	16.1
T27-10D	*ras1*$^-$ *ras2*$^-$ *bcy1*	36.8	0.4	0.4
Experiment 2[b]				
T27-10D	*ras1*$^-$ *ras2*$^-$ *bcy1*	51.0	1.1	0.7
AM18-5C	*RAS1 RAS2 cyr1*-1	0.34	0.12	0.18
T27-10D+AM18-5C		29.5	2.3	23.5

[a] Membranes from the indicated strains were prepared and adenylate cyclase was assayed as described in Ref. *9*. Membranes were assayed either in the presence of 2.5 m*M* Mn^{2+}, or 2.5 m*M* Mg^{2+}, or 2.5 m*M* Mg^{2+} and 10 μ*M* of the nonhydrolyzable GTP analog Gpp(NH)p. Adenylate cyclase activity is expressed in units of picomoles of cAMP generated per milligram of membrane protein per minute. Essentially identical results were obtained in three independent experiments.

[b] In this experiment membranes from the indicated strains were incubated either alone or together for 2 hr at 0°C in 25 m*M* MES, pH 6.2, 1 m*M* ATP, and 0.06% Lubrol with or without 30 μ*M* Gpp(NH)p. They were diluted such that the final Lubrol concentration was 0.01%. Membranes were then incubated at 15°C for 60 min and assayed as before. Essentially identical results were obtained in three independent experiments. See Ref. *9* for details.

ments. We prepared membranes from *RAS1 RAS2* yeast carrying the *cyr1*-1 mutation, and membranes from *ras1*$^{-}$ *ras2*$^{-}$ yeast. The *cyr1*-1 mutation disrupts the catalytic subunit of adenylate cyclase. Membranes from these two sources were assayed separately and together after membrane mixing and fusion. The data (Table VI) indicate that membrane mixing and fusion regenerates a guanine nucleotide–stimulated adenylate cyclase activity.

VII. Discussion

We have demonstrated that the *RAS* genes in yeast are required for growth and viability. Expression of the intact mammalian *H-ras* protein is sufficient to maintain yeast growth and viability in the absence of endogenous *RAS* function. Thus the immediate biochemical function of *ras* proteins has been conserved since the time the progenitors of yeast and mammals diverged during their evolution.

Examination of the phenotype of yeast cells carrying *RAS* mutants suggest that *RAS* is involved in the cAMP pathway. The biochemical experiments presented here indicate that mutants in *RAS* have altered adenylate cyclase activity. Indeed, preliminary results in our lab indicate that purified yeast *RAS* or mammalian *ras* proteins restore guanine nucleotide–stimulated adenylate cyclase activity in membranes prepared from yeast cells lacking endogenous *RAS* proteins.

These results suggest that *ras* proteins may modulate adenylate cyclase in mammalian cells. This is an attractive possibility since changes in cAMP affect cell cycle progression (*7, 19*), and aberrations in the control of adenylate cyclase are seen in many transformed cells (*15*). However, the regulation of adenylate cyclase in mammals is likely to be more complex than in yeast. At least two other guanine nucleotide–binding proteins are known to modulate adenylate cyclase (*6*).

It is important to keep in mind that we do not know how *ras* proteins stimulate adenylate cyclase in yeast. If the mechanism is indirect, it is possible that the chain of interactions has not been maintained intact throughout evolution. In that case, *ras* proteins might not influence adenylate cyclase at all in mammals, or adenylate cyclase might not be the prime target. However, it is likely that a study of the immediate molecular interactions of *RAS* proteins in yeast will provide important clues for the normal function of *ras* in mammalian cells.

References

1. D. Capon, P. Seeburg, J. McGrath, J. Hayflick, U. Edman, A. Levinson, and D. Goeddel, *Nature (London)* **304,** 507–513 (1983).
2. G. Casperson, N. Walker, A. Brasier, and H. Bourne, *J. Biol. Chem.* **258,** 7911–7914 (1983).
3. D. Defeo-Jones, E. Scolnick, R. Koller, and R. Dhar, *Nature (London)* **306,** 707–709 (1983).
4. R. Dhar, A. Nieto, R. Koller, D. Defeo-Jones, and E. Scolnick, *Nucleic Acids Res.* **12,** 3611–3618 (1984).
5. R. Ellis, D. DeFeo, T. Shih, M. Gonda, H. Young, N. Tsuchida, D. Lowy, and E. Scolnick, *Nature (London)* **292,** 506–511 (1981).
6. A. Gilman *Cell* **36,** 577–579 (1984).
7. H. Green, *Cell* **15,** 801–811 (1978).
8. T. Kataoka, S. Powers, C. McGill, O. Fasano, J. Strathern, J. Broach, and M. Wigler, *Cell* **37,** 437–445 (1984).
9. T. Kataoka, S. Powers, S. Cameron, O. Fasano, M. Goldfarb, J. Broach, and M. Wigler, *Cell* in press (1985).
10. K. Matsumoto, I. Uno, Y. Oshima, and T. Ishikawa, *Proc. Natl. Acad. Sci. U.S.A.* **79,** 2355–2359 (1982).
11. K. Matsumoto, I. Uno, and T. Ishikawa *Cell* **32,** 417–423 (1983).
12. K. Matsumoto, I. Uno, and T. Ishikawa, *Exp. Cell Res.* **146,** 151–161 (1983).
13. J. McGrath, D. Capon, D. Goeddel, and A. Levinson, *Nature (London)* **310,** 644–655 (1984).
14. A. Papageorge, D. Lowy, and E. Scolnick, *J. Virol.* **44,** 509–519 (1982).
15. I. Pastan, G. Johnson, and W. Anderson, *Annu. Rev. Biochem.* **44,** 491–522 (1975).
16. S. Powers, T. Kataoka, O. Fasano, M. Goldfarb, J. Strathern, J. Broach, and M. Wigler, *Cell* **36,** 607–612 (1984).
17. E. Reddy, R. Reynolds, E. Santos, and M. Barbacid, *Nature (London)* **300,** 149–152 (1982).
18. C. Rivin and W. Fangman, *J. Cell Biol.* **85,** 96–107 (1980).
19. E. Rozengurt, *in* "Advances in Cyclic Nucleotide Research" (J. Dumont, P. Greengard, and G. Robison, eds.), Vol. 14, pp. 429–442. Raven, New York, 1981.
20. T. Shih, A. Papageorge, P. Stokes, M. Weeks, and E. Scolnick, *Nature (London)* **287,** 686–691 (1980).
21. T. Shih, P. Stokes, G. Smythes, R. Dhar, and S. Oroszian, *J. Biol. Chem.* **257,** 11767–11773 (1982).
22. T. Shih, M. Weeks, P. Gruss, R. Dhar, S. Oroszlan, and E. Scolnick, *J. Virol.* **42,** 253–261 (1982).
23. K. Shimizu, D. Birnbaum, M. Ruley, O. Fasano, Y. Suard, L. Edlund, E. Taparowsky, M. Goldfarb, and M. Wigler, *Nature (London)* **304,** 497–500 (1983).
24. R. Sweet, S. Yokoyama, T. Kamata, J. Feramisco, M. Rosenberg, and M. Gross, *Nature (London)* **311,** 273–275 (1984).
25. C. Tabin, S. Bradley, C. Bargmann, R. Weinberg, A. Papageorge, E. Scolnick, R. Dhar, D. Lowy, and E. Chang, *Nature (London)* **300,** 143–148 (1982).
26. F. Tamanoi, M. Walsh, T. Kataoka, and M. Wigler *Proc. Natl. Acad. Sci. U.S.A.* **81,** 6924–6928 (1984).
27. E. Taparowsky, K. Shimizu, M. Goldfarb, and M. Wigler, *Cell* **34,** 581–586 (1983).
28. E. Taparowsky, Y. Suard, O. Fasano, K. Shimizu, M. Goldfarb, and M. Wigler, *Nature (London)* **300,** 762–765 (1982).

29. K. Tatchell, D. Chaleff, D. Defeo-Jones, and E. Scolnick, *Nature (London)* **309,** 523–527 (1984).
30. T. Toda, I. Uno, T. Ishikawa, S. Powers, T. Kataoka, D. Broek, S. Cameron, J. Broach, K. Matsumoto, and M. Wigler, *Cell* in press (1985).
31. I. Uno, K. Matsumoto, K. Aduchi, and T. Ishikawa, *J. Biol. Chem.* **258,** 10867–10872 (1983).
32. I. Uno, K. Matsumoto, and T. Ishikawa, *J. Biol. Chem.* **257,** 14110–14115 (1982).
33. Y. Yuasa, S. Srivastava, C. Dunn, J. Rhim, E. Reddy, and S. Aaronson, *Nature (London)* **303,** 775–779 (1983).

12

The *neu* Oncogene Encodes a Cell Surface Protein with Properties of a Growth Factor Receptor

DAVID F. STERN, ALAN SCHECHTER,
LALITHA VAIDYANATHAN, ROBERT WEINBERG

Whitehead Institute for Biomedical Research
Massachusetts Institute of Technology
Cambridge, Massachusetts

MARK GREENE AND JEFFREY DREBIN

Department of Rheumatology
Tufts University School of Medicine
Boston, Massachusetts

A large group of cellular oncogenes have been uncovered in the past decade. These oncogenes encode proteins that are found in various intracellular locations, where they affect cellular metabolism in a number of distinct ways. One group of oncogenes has been found by use of gene transfer procedures. These oncogenes may be found in the DNAs of a variety of chemically induced mouse sarcomas as well as in human tumors of spontaneous origin (reviewed in Refs. *1* and *2*).

The use of gene transfer (transfection) in our own laboratory during the period 1978–1980 uncovered several of these oncogenes (*3*, *4*). Because the molecular nature of these genes was unknown, we undertook two lines of experimentation.

We attempted isolation of these genes by molecular cloning and identification of oncogene-encoded proteins by use of immunological techniques.

The immunological approach depended on the fact that the oncogenes under study were able to impart the tumorigenic phenotype to NIH3T3 mouse fibroblasts. We took NIH3T3 cells carrying introduced oncogenes and used these to seed fibrosarcomas in quasi-syngeneic NFS mice. We reasoned that the introduced oncogenes would induce synthesis of novel proteins in the tumor cells and anticipated that these proteins would provoke an immune response in the tumor-bearing mice. Sera from these mice were tested for their ability to precipitate proteins from the transfected cells that were not present in untransfected NIH3T3 cells.

These experiments were uniformly unsuccessful, with one exception. It is clear in retrospect that the bulk of the cells that were analyzed carried oncogenes that are members of the *ras* gene family. These *ras* oncogenes induce synthesis of 21,000-dalton proteins that are highly conserved evolutionarily and located on the cytoplasmic face of the plasma membrane. The single successful experiment stemmed from work on a group of oncogenes found in rat neuroblastoma and glioblastoma cell lines.

The rat tumor cell lines had been produced by others who had exposed pregnant BDIX rats to ethyl nitrosourea (ENU). The rat pups born from this pregnancy displayed neuroectodermal tumors 3–6 months postpartum (*5*). These tumors were used, in turn, to create established cell lines, the DNAs of which we analyzed by transfection (*4*).

The oncogenes present in four independently derived tumor cell lines were closely related to one another in their structure and behavior. Consequently, these oncogenes will be treated collectively here. We have termed them *neu* to reflect their association with the neuroblastomas.

The sera from mice carrying *neu*-induced tumors exhibited a specific and novel reactivity: a 185,000-dalton protein was recognized in *neu*-transfected cells that was absent in their untransfected NIH3T3 counterparts. This p185 protein was present in all NIH3T3 cells transformed by the *neu* oncogene and in none of the NIH3T3 lines carrying a variety of other oncogenes, including oncogenes of the *ras* gene family. We concluded that the display of p185 was a specific consequence of the expression of the *neu* oncogene and did not simply result from the general process of cell transformation (*4*). Animals carrying various *ras*-induced tumors yielded sera having no reactivity with p185, further establishing the specific association of p185 with the *neu* oncogene. Unfortunately, none of these experiments could show that the *neu* oncogene encoded the structure of p185. It remained equally likely that p185 was an NIH3T3-encoded protein whose synthesis was specifically induced by the *neu* oncogene.

Biochemical analysis of the p185 revealed a number of properties that further distinguished it from the *ras*-encoded proteins. Treatment of *neu*-transfected cells with various proteases abolished the reactivity of the p185 with the anti-tumor sera. Because such protease treatment did not affect cell viability and left the bulk of cellular proteins intact, we concluded that p185 possesses extra-cellular domains.

The p185 protein could also be labeled by incubation of cells with inorganic phosphate. Moreover, the electrophoretic mobility of the protein increased when it was synthesized in the presence of tunicamycin. This indicated that p185, like many cell surface proteins, is a glycoprotein.

A clue to the nature of *neu* and p185 was provided by the discovery in 1983 of strong homology between the *erbB* oncogene and the cellular gene encoding the epidermal growth factor receptor (EGFr) (*6*). This established the precedent that an oncogene can be created by alteration of a cellular growth factor receptor gene. The structural similarities between the EGFr and p185 protein led us to the speculation that the *neu* gene may share sequence homology with *erbB*.

DNAs were prepared from a number of NIH3T3 lines carrying transfected *neu* oncogenes. These DNAs were analyzed by the Southern procedure, using an *erbB* oncogene clone as probe. These Southern blots showed that all cells transfected with the *neu* oncogene had concomitantly acquired a novel DNA segment reactive with the *erbB* oncogene probe. This proved that the two genes, *erbB* and *neu,* were either homologous or adjacent and resident on the same DNA fragment (*10*). The latter possibility could be dismissed, because linkage between the two genes survived a large number of transfection events, some of which would be expected to dissociate adjacent genes.

The homology between *erbB* and *neu* further suggested homology between the EGFr and p185. Use of a polyclonal anti-EGFr serum showed some cross-reactivity with p185 (*10*). This led us to believe that EGFr and p185 are structurally related and that p185 is indeed encoded by the *neu* oncogene.

Such data might be interpreted to indicate an identity between p185 and the EGFr. However, more detailed analysis revealed a more complex relationship. This further analysis depended upon development of a monoclonal antibody having specificity for p185. A group of such antibodies was developed using spleens of mice carrying *neu*-induced tumors. One of these, termed 7.16.4, was used in subsequent experiments because it could be used to immunoprecipitate p185. Moreover, this antibody could be used in conjunction with a fluorescence-activated cell sorter to demonstrate specific staining of *neu* transfectants as well as of the neuro-glioblastoma cells from which *neu* originates (*7*).

The monoclonal 7.16.4 antibody was absolutely specific in its recognition of p185 and showed no reactivity with the human or rat EGFr species. In contrast, the polyclonal anti-EGFr serum showed much stronger reactivity with the EGFr than with p185. These results suggested that the two proteins, while related, were distinct in at least a portion of their epitopes.

A more striking difference emerged from study of the electrophoretic mobilities of the two proteins. EGFr migrated with a mobility compatible with a protein of 170,000 daltons, this being about 15,000 less than the apparent mass of p185. This difference in mobility continued to be observed when the two proteins were synthesized in the presence of tunicamycin. Such result indicates that the differences between the two proteins can not be attributed to differential N-linked glycosylation.

These results show that the *neu*-encoded p185 and the EGFr are related proteins having distinct structures. Such differences in structure might derive from two sources. It may be that the p185 is encoded by the same gene that specifies the EFGr. In this instance, the difference in their molecular masses may be caused by the same genetic lesions that led to oncogenic activation of the antecedent protooncogene.

The alternative possibility is that the *neu* oncogene stems from a protooncogene that is separate and distinct from the EGFr gene. Such a proto-*neu* may have homologies with the EGFr but encode the structure of a distinct protein. Currently available evidence militates toward this second alternative. Thus, use of the 7.16.4 anti-p185 monoclonal antibody has revealed a 182,000-dalton protein that is present in low amounts in untransformed cells of the Rat-1 fibroblast cell line and is distinct from 170-kdal EGFr in these cells (*10*). Because of the specificity of the 7.16.4 antibody, it appears likely that this normal 182-kdal protein is antigenically and structurally related to the oncogene-encoded p185 and that its structure is specified by the *neu* protooncogene. Thus, the *neu* protooncogene appears to be distinct from the normal EGFr gene.

It is thus possible that the p185 oncogene protein differs only minimally in structure from its normal counterpart. In accordance with this is the fact that the carcinogen responsible for activation of the *neu* gene was ENU, whose activity favors the creation of point mutations. Perhaps alterations in p185 amino acid sequence are important to its oncogenic activity. The oncogene-encoded protein is able to induce cellular transformation when present in low amounts. This suggests that a genetic lesion resulting in deregulation of expression was not responsible for activation of this gene.

The p185 appears to represent an altered form of a cellular growth factor receptor that is distinct from the EGFr. It may represent the receptor for a known growth factor, such as transforming growth factor alpha (TGFα) or platelet-derived growth factor (PDGF). Alternatively, it may represent the gene encoding the receptor of an as yet unidentified growth factor.

All of this adds to the results already developed by others who have implicated cellular genes that encode growth factors and growth factor receptors as targets for mutations leading to oncogenic activation (*6, 8*). Taken together, the results cause one to return to a large body of data which indicates that tumor cells differ from normal cells in their relative independence from exogenous growth factors. In this context, oncogenes may be viewed as acting by conferring on cells an ability to grow in spite of the absence of normally required factors. Some oncogenes may confer this autonomy by allowing the tumor cell to make its own growth stimulatory factors, thus creating an "autocrine" or positive feedback loop of growth stimulatory factors (*9*). In fact, the *sis* oncogene is known to directly encode such a factor, PDGF (*8*).

An alternative mechanism of factor autonomy stems from the work on *erbB* and the EGFr, as well as the present results. Here one presumes that autonomy is achieved by altering the proteins that the cell uses to sense the presence of factors in the extracellular space. These alterations may lead to receptors that are constitutively activated, seemingly informing the cell about the presence of such factors, even when they may be absent. Detailed structural analysis of the *neu* oncogene, presently underway, should rapidly produce evidence to support these conjectures.

Unrelated to this, but equally intriguing, is the role that the p185 plays as a cell surface tumor antigen. In contrast to almost all the other oncogene-encoded proteins, the p185 is displayed on the cell surface. Its continued presence may be required for maintenance of the transformed state of the tumor cell. This contrasts to a number of other tumor cell surface antigens, whose display is probably gratuitous to the growth properties of the tumor cell. Expression of these other antigens may thus be modulated without loss of growth impetus; in contrast, loss of oncogenic p185 may cause a transformed cell to revert to normal phenotype. This logic leads us to consider p185 as a good target for recognition by antibodies that, having recognized this protein, prevent growth of the *neu*-transformed cell. We are currently studying the effects of anti-p185 on *neu* transformants growing *in vitro* and as tumors in mice. This may yield a useful experimental model by which we can explore the interactions of the immune system with tumors bearing a specific cell surface antigen.

Acknowledgments

We thank Patricia Heffernan for excellent technical assistance. This work was supported by grants from the American Business Cancer Foundation as well as Grant No. CA 26717 of the U.S. National Cancer Institute.

References

1. J. M. Bishop, *Ann. Rev. Biochem.* **52,** 301–354 (1983).
2. H. Land, L. Parada, and R. A. Weinberg, *Science* **222,** 771–778 (1983).
3. C. Shih, L. C. Padhy, M. Murray, and R. A. Weinberg, *Nature (London)* **290,** 261–264 (1981).
4. L. D. Padhy, C. Shih, D. Cowing, R. Finkelstein, and R. A. Weinberg, *Cell* **28,** 865–871 (1982).
5. D. Schubert, *Nature (London)* **249,** 224 (1974).
6. J. Downward *Nature (London)* **307,** 521–527 (1984).
7. J. A. Drebin, D. F. Stern, V. C. Link, R. A. Weinberg, and M. I. Greene, *Nature (London)* **312,** 545–548 (1984).
8. M. D. Waterfield, *Nature (London)* **304,** 35–39 (1983).
9. M. B. Sporn and G. J. Todaro *N. Engl. J. Med.* **303,** 878–880 (1980).
10. A. L. Schechter, D. F. Stern, L. Vaidyanathan, S. J. Decker, J. A. Drebin, M. I. Greene, and R. A. Weinberg, *Nature (London)* **312,** 513–516 (1984).

13

Cooperativity between "Primary" and "Auxiliary" Oncogenes of Defective Avian Leukemia Viruses

THOMAS GRAF, BECKY ADKINS, ACHIM LEUTZ, HARTMUT BEUG, AND PATRICIA KAHN

European Molecular Biology Laboratory
Heidelberg, Federal Republic of Germany

I. Introduction

The avian sarcoma viruses and defective leukemia viruses (DLVs) are acutely transforming retroviruses that contain cell-derived oncogenic sequences (*1*, *2*). The DLVs cause predominantly hematopoietic neoplasms in chicks and transform bone marrow cells in culture, with most viral strains showing specificity for cells within one hematopoietic lineage (*3*). As shown in Table I, DLVs can be divided into three major groups:

GENETICS, CELL DIFFERENTIATION, AND CANCER

TABLE I

Defective Avian Leukemia Viruses

Virus	Oncogene(s)	Hematopoietic neoplasms	*In vitro* transformed hematopoietic cells
AEV-H	*erbB*	Erythroblastosis	Erythroblasts
AEV-ES4	*erbA, erbB*	Erythroblastosis	Erythroblasts
AMV	*myb*	Myeloblastosis	Myeloblasts
E26	*myb ets*	Myeloblastosis, erythroblastosis	Myeloblasts, erythroblasts
MC29	*myc*	Myelocytomatosis	Macrophages
OK10	*myc*	?	Macrophages
CMII	*myc*	Myelocytomatosis	Macrophages
MH2	*myc, mil*	Macrophage tumors	Macrophages

1. The *erbB*-containing avian erythroblastosis viruses (AEVs) transform erythroid cells *in vivo* and *in vitro* and are capable of inducing sarcomas and of transforming fibroblasts.
2. The *myb*-containing viruses AMV and E26 transform myeloblasts *in vivo* and *in vitro*. The E26 strain also induces erythroblastosis (the predominant type of leukemia caused by this strain) and transforms erythroblasts and fibroblasts *in vitro* (*5, 6*).
3. The *myc*-containing viruses cause myelocytomatosis (neoplasms consisting of granulocyte–macrophage–precursor cells) as well as a variety of non-hematopoietic neoplasms and transform both macrophages and fibroblasts *in vitro*.

Three DLV strains, one in each group, contain a second, cell-derived, putative oncogene sequence. The ES4 or R strain of AEV contains the *erbA* gene in addition to *erbB* (*9*), the E26 strain contains the *ets* gene in addition to *myb* (*10, 11*), and the MH2 strain contains the *mil* gene in addition to *myc* (*12, 13*). The studies described in this paper focus on the question of what effects these putative second oncogenes have on the transformation of avian hematopoietic cells.

II. The *erbA* Gene Induces a Tight Block of Differentiation in *erbB*-Transformed Erythroblasts

Previous studies with the *erbA, erbB*–carrying AEV-ES4 strain have shown that *erbB* alone is sufficient to cause both erythroblastosis and sarcomas in chicks, while *erbA* by itself has no apparent effect on the properties of infected cells (*14, 15*). However, there are clear phenotypic differences between erythroid cells transformed with an ES4 mutant lacking the *erbA* gene and those transformed with wild-type virus containing both genes: the former tend to differentiate spontaneously into reticulocyte- and erythrocyte-like cells, while the latter rarely do so (*14*). Furthermore, *erbB* erythroblasts have more stringent growth requirements than do *erbA, erbB*–transformed erythroblasts. These results suggest that the *erbA* gene confers a more transformed phenotype upon erythroblasts transformed with *erbB*. The observed cooperativity seems to be restricted to erythroid cells, since no clear phenotypic differences were detected between fibroblasts transformed with *erbB* only and those transformed with both *erbA* and *erbB* (*14, 16*).

III. The *ets* Gene Appears to Induce a Tight Block of Differentiation in *myb*-Transformed Myeloblasts

The fact that E26 virus (*myb, ets*) differs from AMV (*myb*) in its ability to transform erythroid and fibroblastic cells in addition to myeloblasts suggests that the *myb* gene is responsible for myeloblast transformation and the *ets* gene for erythroblast and fibroblast transformation. This possibility is supported by our recent finding that mutants in E26 virus that are thermolabile for the transformation of myeloblasts behave like wild-type virus with regard to their erythroblast- and fibroblast-transforming capacities (*16*).

To determine whether *ets* has an effect on the phenotype of *myb*-transformed myeloblasts, we compared AMV- and E26-transformed myeloblasts with respect to their expression of macrophage differentiation markers [cellular morphology, adherence, and phagocytic capacities as well as expression of myeloblast- and macrophage-specific cell surface antigens detected with monoclonal antibodies (*17*)]. As summarized in Table II, AMV-transformed myeloblasts tend to differentiate spontaneously into macrophagelike cells, while E26 cells do not. The proportion of differentiated cells varies between individual clones of AMV-transformed myeloblasts and increases with time in culture. In addition, the tumor-promoting phorbol ester TPA efficiently induces AMV- but not E26-transformed myeloblasts to differentiate (*16, 18, 19*). These results suggest that the *ets* gene blocks the TPA-induced differentiation in *myb*-transformed myeloblasts, although it has not yet been ruled out that the *myb* genes carried by the

TABLE II

Differentiation Capacities of Myeloblasts Transformed by AMV and E26 Leukemia Viruses

Infecting virus	Spontaneous differentiation	Inducibility of differentiation by TPA
AMV	+	+
E26	−	−

two viral strains differ slightly. Additional studies using *myb*-only and *ets*-only mutants of E26 virus will be required to clarify this point.

IV. *src*-Type Oncogenes Stimulate cMGF-Independent Growth in *myb,ets*-Transformed Myeloid Cells

Myeloid cells transformed by *myb*-, *myb,ets*-, or *myc*-containing viruses require a specific growth factor for their proliferation (Table III and Fig. 1). This factor, termed cMGF for chicken myelomonocytic growth factor, also stimulates the production of macrophage and probably granulocyte colonies in normal bone marrow cultures (*20*). cMGF is obtained either from medium conditioned by concanavalin-A (ConA) –treated chick spleen cells (*21*) or from an MC29 virus–transformed macrophage cell line following treatment with LPS (*20*). cMGF from the latter source has been purified to homogeneity, and antibodies were produced against it in rabbits (*20*).

MH2 is exceptional among the DLVs with myeloid transforming capacity in that it is the only strain which yields cMGF-independent transformed myeloid cells. Since this strain also contains the *src*-related *mil* gene in addition to *myc*, we have investigated whether *src*-type oncogenes play a role in inducing factor independence in transformed chicken myeloid cells.

To determine whether oncogenes other than *mil* can induce cMGF independence in different cMGF-dependent myeloid cell types, we superinfected factor-dependent E26 myeloblasts with various oncogene-carrying retroviruses and assayed the cells several days later for their cMGF requirement.* The results, which

*E26 cells were chosen rather than AMV cells because they are easier to propagate as a homogeneous population.

TABLE III

cMGF Dependence of *myb*- and *myc*-Transformed Myelomonocytic Cells

Transforming virus	Oncogene(s)	Phenotype of transformed cells	cMGF dependence
AMV	*myb*	Myeloblasts	Yes
E26	*myb, ets*	Myeloblasts	Yes
MC29	*myc*	Macrophages	Yes
OK10	*myc*	Macrophages	Yes
CMII	*myc*	Macrophages	Yes
MH2	*myc, mil*	Macrophages	No[a]

[a] A low proportion of transformed macrophage clones is cMGF dependent.

are summarized in Table IV, indicate that viruses containing the *src, fps, yes, erbB, mil,* or *ros,* but not the *myc* oncogene, render E26-transformed myeloblasts growth factor independent. We then tested whether the superinfected cells produce their own factor. As shown in Fig. 2, cells superinfected with either a *src, fps, erbB,* or *yes*-containing virus and grown in serum-free medium secrete an activity that stimulates the proliferation of unsuperinfected E26 my-

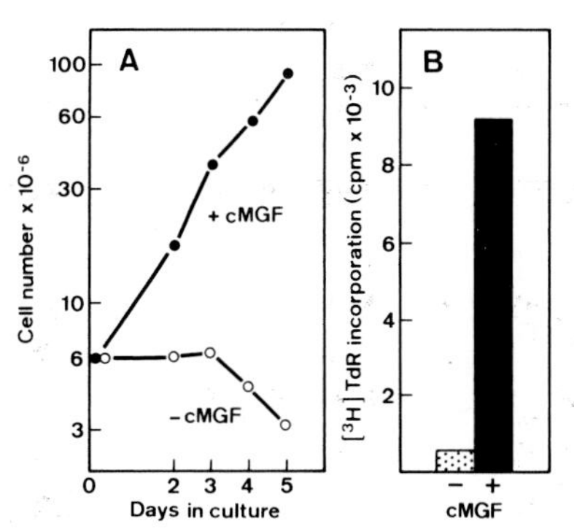

Fig. 1. Assay of cMGF dependence of E26-transformed myeloblasts. (A) Transformed myeloblasts seeded at 37°C in growth medium with 1% (+) or without (−) ConA-stimulated spleen cell–conditioned medium as a source for cMGF. Cells were fed every 2 days and cell numbers determined at daily intervals. (B) E26 myeloblasts (2×10^4) were seeded in 50 μl medium with or without cMGF, as above. Two days later, samples were labeled for 2 hr with [^{3}H]thymidine and incorporated radioactivity determined.

TABLE IV

Induction of cMGF Independence in E26-Transformed Myeloblasts by Superinfection with Viruses Containing *src*-Type Oncogenes

Oncogene	Virus strain	[^{3}H]TdR incorporation −cMGF/+cMGF
src	SR-RSV-D	0.72
fps	PRCII	0.83
yes	Y73	1.17
erbB	AEV-H	1.47
ros	UR2	1.13
myc	MC29	0.04
—	RAV-2	0.06
—	—	0.08

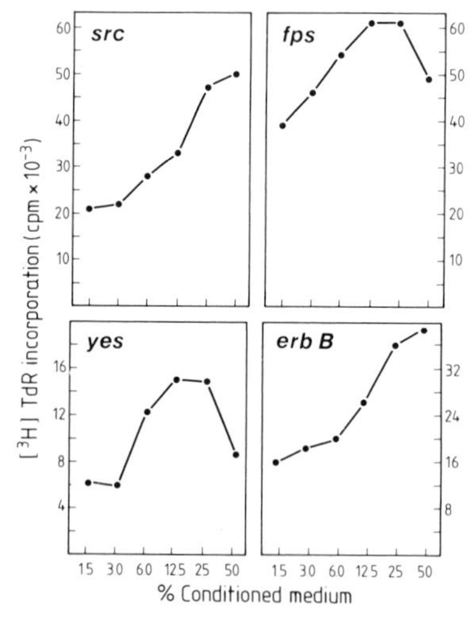

Fig. 2. E26 myeloblasts superinfected with viruses containing *src*-type oncogenes produce a myeloblast growth–stimulating activity. Superinfected cells were washed and incubated for 16–18 hr in serum-free medium. Conditioned medium was obtained by removing cells and viruses by low- and high-speed centrifugation and tested in different concentrations on unsuperinfected E26 myeloblasts. The oncogenes indicated were introduced using the virus strains listed in Table IV.

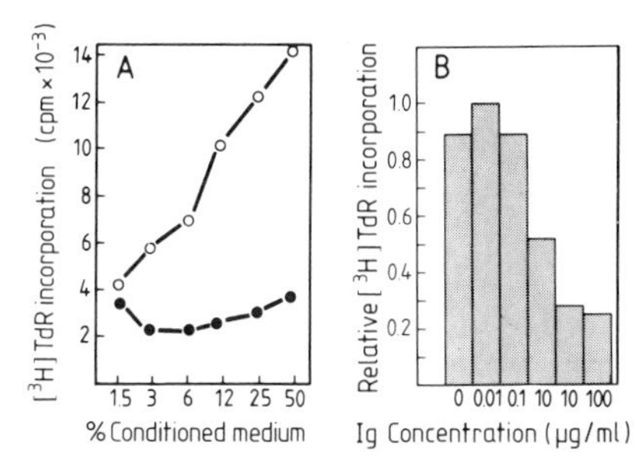

Fig. 3. Superinfected E26 myeloblasts are dependent on a cMGF-like growth factor that they themselves produce. (A) Conditioned medium prepared from Y73 virus–superinfected E26 myeloblasts was incubated for 16 hr at 4°C with rabbit anti-cMGF (●) or preimmune serum (○). The samples were then clarified by centrifugation and tested for growth factor activity. (B) Same cells as in A were incubated in microtiter wells with increasing concentrations of immunoglobulin (Ig) fractions from anti-cMGF serum or preimmune Ig. Antibodies were added again on day 1 and thymidine incorporation was determined on day 2. The values plotted indicate the ratio of [^{3}H]TdR incorporation of cells treated with anti-cMGF Ig and preimmune Ig.

eloblasts. However, preincubation of the conditioned medium from *yes*-superinfected E26 cells with antibodies to cMGF completely neutralized the stimulatory activity (Fig. 3A). Similar results were also obtained with conditioned media from *fps*- and *erbB*-superinfected E26 cells (data not shown). Furthermore, incubation of *yes*-superinfected E26 cells with anti-cMGF antibodies for 2 days resulted in a nearly complete inhibition of growth (Fig. 3B). The inhibitory effect of the antibodies could be overcome by the addition of excess cMGF, ruling out the possibility that the antiserum was toxic to the cells. These results show that E26 myeloblasts superinfected with viruses containing *src*-type oncogenes are stimulated to secrete cMGF or a cMGF-like factor which they utilize for their own growth. Studies using *ts* mutants of RSV indicate that this induction is dependent on the continuous synthesis of an active $p60^{src}$ protein (*22*).

Figure 4 shows a scheme which summarizes our data on the oncogene cooperativity in *myb*-transformed myeloblasts. The combination of the *myb* and *ets* oncogenes induces a complete block of differentiation without affecting cMGF dependence, while the introduction of a *src* family oncogene into these cells results in cMGF independence but not in other detectable changes. It is unlikely that *ets* has a role in this latter phenomenon, since preliminary data indicate that AMV-transformed myeloblasts superinfected with viruses carrying *src*-type on-

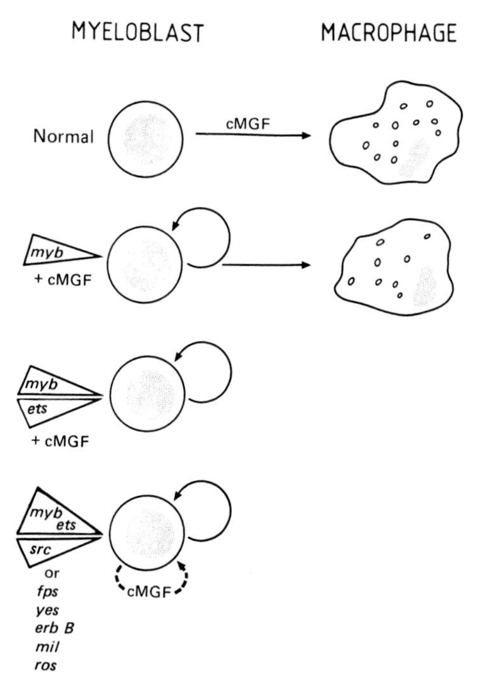

Fig. 4. Diagram depicting cooperativity between *myb* and *ets* genes and between *myb*, *ets* and *src*-type genes in transformed myeloblasts. Straight arrows, differentiation; curved arrows, self renewal; broken arrows, production of growth factor.

cogenes also lose their cMGF requirement (B. Adkins and T. Graf, unpublished observations).

V. *src*-Type Oncogenes Abolish the cMGF Requirement of *myc*-Transformed Macrophages

Superinfection of macrophages transformed by the *myc*-only–containing virus OK10 (or MC29) with viruses carrying the *fps, erbB, mil,* or (to a lesser extent) *yes* oncogene causes them to become cMGF independent for growth. For reasons that are unclear, *src*-superinfected OK10 cells remain cMGF dependent, even when different strains of RSV are used (Table V) (*22*).

The MH2 virus probably represents a natural case of cooperativity between *myc*- and *src*-type oncogenes: preliminary evidence shows that mutants of MH2

TABLE V

Induction of cMGF Independence in OK10-Transformed Macrophages by Superinfection with Viruses Containing *src*-Type Oncogenes

Oncogene	Virus strain	[^{3}H]TdR incorporation −cMGF/+cMGF
src	SR-RSV-D	0.10
fps	PRCII	3.60
yes	Y73	0.51
erbB	AEV-H	2.40
mil, myc	MH2	1.67
myb, ets	E26	0.12
myb	AMV	0.01
—	—	0.06

lacking the *mil* gene still transform macrophages but that these cells are completely cMGF dependent. *In vivo,* wild-type MH2 virus induces predominantly macrophage-type neoplasms consisting of factor-independent cells, while a deletion mutant lacking part of the *mil* gene is considerably less pathogenic and induces mostly neoplasms of nonhematopoietic origin (Graf *et al.*, in preparation). These findings suggest that MH2 has evolved as a highly pathogenic strain from a *myc*-containing virus that incorporated the *mil* gene into its genome. A scheme summarizing results on oncogene cooperativity in macrophagelike cells is shown in Fig. 5.

VI. Concluding Remarks

Our studies have shown that several acute leukemia virus isolates which contain two cell-derived oncogenic sequences show a cooperation between their oncogenes within a specific hematopoietic lineage. Each of the strains which we have examined contains what could be called a "primary" oncogene that transforms hematopoietic cells of a given lineage, and an "auxiliary" oncogene that induces a more highly transformed phenotype in these already transformed cells (Table VI). In case of the AEV-ES4 strain the primary oncogene is *erbB,* which corresponds to a truncated form of the EGF receptor (*23*) and which may act as a constitutively switched-on growth factor receptor. The phenotype of *erbB*-transformed erythroblasts is enhanced by the *erbA* gene in that the presence of this

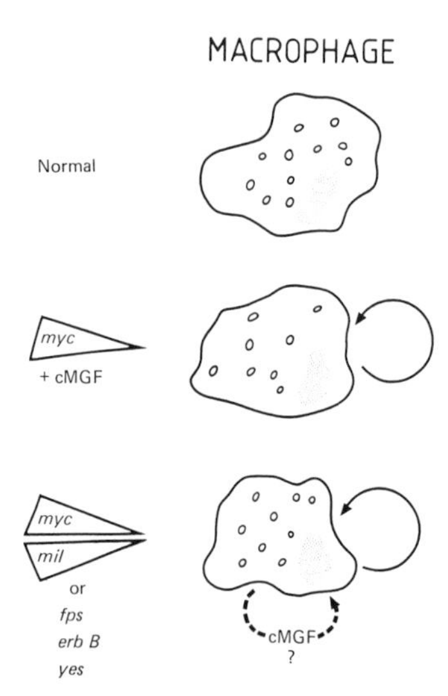

Fig. 5. Diagram depicting cooperativity between *myc* and *src*-type genes in transformed macrophages. Curved arrows, self-renewal; broken arrows, growth factor production.

second oncogene leads to a complete block of differentiation and to a reduction in the complexity of their growth medium requirements. The *in vivo* effects of *erbA* are less well studied, although initial experiments comparing AEVs with *erbB* only to those with *erbA* and *erbB* suggest that virus containing both oncogenes is more leukemogenic (*14*).

We are currently investigating whether *erbA* can enhance transformation in

TABLE VI

Cooperativity of Viral Oncogenes in Different Types of Hematopoietic cells

Virus	Primary oncogene	Auxiliary oncogene	Type of cell(s) in which cooperativity is manifested
AEV-ES4	*erbB*	*erbA*	Erythroblasts
E26	*myb*	*ets*	Myeloblasts, erythroblasts?
MH2	*myc*	*mil*	Macrophages

erythroid cells transformed by oncogenes other than *erbB*. The experiments are being carried out using erythroblasts transformed by the *src* oncogene of Rous sarcoma virus; these cells have properties similar to erythroblasts transformed by *erbB* only (*28*). Preliminary results suggest that erythroblasts transformed with both *src* and *erbA* have a phenotype resembling *erbA*, *erbB*-transformed erythroblasts and that *erbA* can shift the pathogenicity of Rous sarcoma virus from a predominantly sarcomagenic agent to that of an acute erythroleukemia virus (*24;* P. Kahn, L. Frykberg, B. Vennström, and T. Graf, in preparation).

The cooperativity between the *myb* and *ets* oncogenes is less well understood. Differentiation into macrophagelike cells appears to be tightly blocked in myeloblasts transformed by *myb, ets* but not in myeloblasts transformed by *myb* alone. It is not yet clear whether *myb* and *ets* also cooperate in the transformation of erythroblasts, the main cell type transformed by E26 *in vivo*. It also remains to be determined whether the ability of *src*-type oncogenes to induce autocrine growth in *myb*-transformed myeloblasts (described in this paper) has any consequences *in vivo*. So far, no natural virus isolates are known that encode both *myb*- and *src*-type oncogenes.

The MH2 strain provides a third example of a virus that carries two cooperating oncogenes. The *myc* gene encodes the macrophage-transforming capacity and thus serves as the primary oncogene, while the *mil* gene renders the *myc*-transformed macrophages cMGF independent. This cooperativity appears to be an important factor contributing to the high pathogenicity of MH2 virus.

Our results also suggest that oncogenes may play different roles in different cell types. For example, *src*-type oncogenes are sufficient to induce full neoplastic transformation in fibroblasts; in erythroid cells they also function as primary oncogenes, but these transformed cells can acquire a more highly transformed phenotype through the action of the auxiliary *erbA* oncogene. Finally, in *myb*- or *myc*-transformed myeloid cells, *src*-type oncogenes act as auxiliary oncogenes by abolishing the growth factor requirement of these cells without being able to directly transform them.

Studies with mammalian cells have shown that *myc* can cooperate with the *ras* oncogene to induce a fully transformed phenotype in primary fibroblasts (*25*). Although the basis of this cooperativity is not yet fully understood, it has been suggested that *myc* acts an immortalizing gene, while *ras* acts as a transforming gene. The fact that essentially all *myc*-transformed chick macrophage and fibroblast clones tested have a limited lifespan (*7, 26*) argues against an immortalizing role for *myc* in avian cells. In contrast, the action of the *ras* gene in rodent cells resembles that of *src* in chicken cells in that *ras* induces the production of growth factors (*27*). In this context it will be interesting to determine whether or not the *ras* gene can also cooperate with *myc* in transformed chick macrophages.

References

1. T. Graf and D. Stehelin, *Biochim. Biophys. Acta Revs. Cancer* **651,** 245 (1982).
2. J. M. Bishop, *Annu. Rev. Biochem.* **52,** 301 (1983).
3. T. Graf and H. Beug, *Biochim. Biophys. Acta Revs. Cancer* **516,** 269 (1978).
4. C. Moscovici, J. Samarut, L. Gazzolo, and M. G. Moscovici, *Virology* **113,** 765 (1981).
5. K. Radke, H. Beug, G. Doederlein, and T. Graf, *Exp. Cell Res.* **143,** 383 (1982).
6. T. Graf, N. Oker-Blom, T. G. Todorov, and H. Beug, *Virology* **99,** 431 (1979).
7. H. Beug, A. V. Kirchbach, G. Doederlein, J.-F. Conscience, and T. Graf, *Cell* **18,** 375 (1979).
8. L. Gazzolo, C. Moscovici, M. G. Moscovici, and J. Samarut, *Cell* **16,** 627 (1979).
9. B. Vennström and J. M. Bishop, *Cell* **28,** 135 (1982).
10. D. Leprince, A. Gegonne, J. Coll, C. deTaisne, A. Schneeberger, C. Lagrou, and D. Stehelin, *Nature (London)* **306,** 391 (1983).
11. F. E. Nunn, P. H. Seeburg, C. Moscovici, and P. H. Duesberg, *Nature (London)* **306,** 391 (1983).
12. J. Coll, M. Righi, C. deTaisne, C. Dissous, A. Gegonne, and D. Stehelin, *EMBO J.* **2,** 2189 (1983).
13. H. W. Jansen, B. Ruckert, R. Lurz, and K. Bister, *EMBO J.* **2,** 1969 (1983).
14. L. Frykberg, S. Palmieri, H. Beug, T. Graf, M. J. Hayman, and B. Vennström, *Cell* **32,** 227 (1983).
15. L. Sealy, M. L. Privalsky, G. Moscovici, C. Moscovici, and J. M. Bishop, *Virology* **130,** 155 (1983).
16. H. Beug, A. Leutz, P. Kahn, and T. Graf, *Cell* **39,** 579–588 (1984).
17. S. Kornfeld, H. Beug, G. Doederlein, and T. Graf, *Exp. Cell Res.* **143,** 383 (1983).
18. S. Pessano, L. Gazzolo, and C. Moscovici, *Microbiologica* **2,** 379 (1979).
19. K. H. Klempnauer, G. Symonds, G. I. Evan, and J. M. Bishop, *Cell* **37,** 537 (1984).
20. A. Leutz, H. Beug, and T. Graf, *EMBO J.* **3,** 3191–3197 (1984).
21. H. Beug, M. J. Hayman, and T. Graf, *EMBO J.* **1,** 1069 (1982).
22. B. Adkins, A. Leutz, and T. Graf, *Cancer Cells* **3,** in press (1985).
23. J. Downward, Y. Yarden, E. Mayes, G. Scarce, N. Totty, P. Stockwell, A. Ullrich, J. Schlessinger, and M. D. Waterfield, *Nature (London)* **307,** 521 (1984).
24. B. Adkins, A. Leutz, and T. Graf, *Cell* **39,** 439–445 (1984).
25. H. Land, L. F. Parada, and R. A. Weinberg, *Science* **222,** 771 (1983).
26. B. Royer-Pokora, H. Beug, M. Claviez, H.-H. Winkhardt, R. R. Friis, and T. Graf, *Cell* **13,** 751 (1978).
27. P. L. Kaplan, M. Anderson, and B. Ozanne, *Proc. Natl. Acad. Sci. U.S.A.* **79,** 485 (1982).
28. P. Kahn, B. Adkins, H. Beug, and T. Graf, *Proc. Natl. Acad. Sci. U.S.A.* **81,** 7122–7126.

14

The Family of Human T-Lymphotropic Retroviruses Called Human T-Cell Leukemia/Lymphoma Virus (HTLV): Their Role in Lymphoid Malignancies and Lymphosuppressive Disorders (AIDS)

R. C. GALLO, L. RATNER, M. POPOVIC, S. Z. SALAHUDDIN, M. G. SARNGADHARAN, F. WONG-STAAL, G. SHAW, B. HAHN, M. REITZ, M. ROBERT-GUROFF

Laboratory of Tumor Cell Biology
National Cancer Institute
Bethesda, Maryland

P. D. MARKHAM

Department of Cell Biology
Litton Bionetics, Inc.
Kensington, Maryland

J. GROOPMAN

Hematology/Oncology Division
Deaconess Medicine
Boston, Massachusetts

B. SAFAI

Memorial Sloan-Kettering Institute
New York, New York

GENETICS, CELL DIFFERENTIATION, AND CANCER

I. Introduction

The human T-cell leukemia (lymphotropic) viruses, collectively called HTLV, comprise a family of retroviruses that have in common their affinity for infection of human T cells, especially the T4 subset. These are the first and to date only known human retroviruses. Three subgroups have been identified. HTLV-I, isolated in 1980 (*1*), has been etiologically linked with adult T-cell leukemia (ATL) (*2*). While a large number of HTLV-I isolates, all apparently nearly identical in nucleotide sequence, have been obtained, variants have also been identified. HTLV-Ib, for example, was isolated from an African ATL patient (*3*). Viruses closely related to HTLV-I have also been isolated from Old World primates (*4, 5*) and have been called primate T-cell leukemia viruses, or PTLV. HTLV-II was first isolated from cells of a patient with T-cell hairy cell leukemia (*6*). Whether an association of this subgroup with hairy cell leukemia exists has not yet been established. HTLV-III is the most recent subgroup identified (*7, 8*), and already more than 100 isolates have been obtained. Evidence for its association with the acquired immunodeficiency syndrome (AIDS) will be one of the primary topics reviewed here. We believe additional HTLV family members will be discovered with properties common to HTLV-I, -II, and -III as outlined in Table I, including (1) T-cell tropism, (2) genomic organization into *gag, pol, env,* and *px* genes with some sequence homology between subgroup members, (3) similar protein profiles including a major core protein of approximately 24,000 molecular weight and a Mg^{2+}-preferring reverse transcriptase, (4) some immunologic cross-reactivity between proteins of different subgroups, and (5)

TABLE I

Common Properties and Relationships of HTLV-I, HTLV-II, and HTLV-III

Property	Subgroup of HTLV		
	I	II	III
General infectivity	Lym	Lym	Lym
Particular tropism	T4	T4	T4
RT size	~100,000	~100,000	~100,000
RT divalent cation	Mg^{2+}	Mg^{2+}	Mg^{2+}
Major core protein	p24	p24	p24
Common envelope epitope	+	+	+
Common p24 epitope	+	+	+
Nucleic acid homology to I (stringent)		±	−
Nucleic acid homology to I (moderate stringency)		++	+
Homology to other retroviruses except PTLV	0	0	0
Genome contains a *pX* region	+	+	+
Produces giant multinucleated cells	+	+	+
African origin	Likely	?	Likely

some degree of cytopathic effect on infected cells ranging from syncytial formation and the creation of giant multinucleated cells to frank cell killing. The intriguing aspects of these viruses concern the mechanism(s) by which they exert their biological effects. Although the viruses share many common features, the known proliferative family members, HTLV-I and -II, cause growth of T cells they infect and often transform them. On the contrary, HTLV-III isolates are cytopathic viruses that kill most lymphoid cells infected *in vitro*. Studies illustrating these disparate effects will be reviewed here together with the recent investigations concerning AIDS. The background studies on HTLV-I and -II will not be extensively covered here. For a recent review see Ref. *9*.

II. Transformation of T Cells by HTLV-I and HTLV-II

HTLV-I has clearly been etiologically linked to ATL. As such it must somehow cause unrestrained proliferation of target T cells. From the very first attempts to transmit HTLV-I to permissive cells, it quickly became evident that the virus was capable of transforming T cells. HTLV-I from a variety of sources was used, either in cocultivation procedures involving fresh human T cells and lethally irradiated HTLV-positive leukemic T-cell lines or cell-free virus preparations, and the virus was shown to be effectively transmitted to cord blood T cells and adult bone marrow lymphocytes (*10–14*). Adult peripheral blood leukocytes were more difficult to infect (*15*). Successful cell transformation was demonstrated by the ability of the infected cells to grow as long-term cell lines in the absence of or in diminished concentrations of T-cell growth factor (TCGF) (also called interleukin 2). Following transformation by HTLV, the properties of the infected T cells were altered as summarized in Table II. These changes included both morphological alterations, primarily as a result of mild cytopathic effects of the virus, and phenotypic changes. The latter included acquisition of new surface antigens, such as additional HLA-A and -B locus antigens, and changes in the density of antigens usually expressed on uninfected T cells, including increased density of TCGF receptors, HLA-DR markers, and transferrin (*16, 17*). The cells that were transformed by HTLV-I appeared to be mature T cells. They expressed typical T-cell marker antigens, including receptors for sheep erythrocytes, and were reactive with standard T-lymphocyte–specific monoclonal antibodies. While infected cells were generally positive for the OKT4/Leu 3a antigen of the T–helper/inducer subset, infected bone marrow cultures often expressed markers typical of the T–suppressor/cytotoxic subset, that is, OKT8/Leu 2a antigen, or else neither phenotypic marker (*18*).

One of the more interesting biological properties of HTLV-I–transformed lymphocytes is their constitutive production of one or more lymphokines, that is, soluble, biologically active factors with potential regulatory activity. Cell lines established directly from the peripheral blood of HTLV-I–positive patients with T-cell malignancies or by *in vitro* transformation of cord blood or bone marrow T cells have been shown to synthesize numerous factors including macrophage migration-enhancing factor, leukocyte migration inhibitory factor, migration-enhancing factor, macrophage-activating factor, differentiation-inducing activity, colony-stimulating factor, eosinophil growth and maturation activity, fibroblast-activating factor, and γ interferon (*19*). Preliminary experiments also suggest that biological activities related to B-cell growth factor and platelet-derived growth factor may also be produced. A few HTLV-I–transformed cell lines also produce TCGF (*20*). These substances, of course, all have the potential to interact with other cells of the immune system and regulate or alter normal function.

While HTLV-I clearly causes T-cell proliferation, several kinds of evidence

TABLE II

Comparison of Phenotypic Characteristics of HTLV-I–Infected Cells with PHA-Stimulated Normal Lymphocytes

Characteristic	PHA-stimulated lymphocytes	HTLV-infected leukemic cells	HTLV-transformed lymphocytes
In vitro growth			
Growth pattern	Single cells; small clumps	Large clumps	Large clumps
Gross morphology	Uniform size; 1 nucleus	Variable size; 1 to several nuclei with convolutions; 1 or 2 nucleoli	Variable size; 1 to several nuclei with convolutions; 1 or 2 nucleoli
TCGF requirement	Yes	None or diminished	None or diminished
New antigen expression			
Tac	2–38% +	50–90% +; increased density/cell	50–90% +; increased density/cell
HLA-DR	2–36% +	60–90% +; decreased density/cell	60–90% +; increased density/cell
4D-12	<1–80% +	8–95% +; increased density/cell	80–95% +; increased density/cell
HLA		Additional HLA-A	Additional HLA-A

(*continued*)

TABLE II (*Continued*)

Characteristic	PHA-stimulated lymphocytes	HTLV-infected leukemic cells	HTLV-transformed lymphocytes
		and B locus antigens	and B locus antigens
HTLV proteins	Negative	p19, p24, RT positive	p19, p24, RT positive
HTLV particles	Negative	Positive	Positive
B-cell markers			
EBNA	Negative	Negative	Negative
BA-1	Negative	Negative	Negative
Surface Ig	Negative	Negative	Negative
T-cell markers			
E-rosette	—	80–95% +	—
TdT	—	Negative	Negative
Thymus T-cell markers (OKT6, NA1/34)	—	Negative	Negative
Pan T (T101, OKT3)	90–98% +	50–100% +	50–100% +
OKT3, Leu 3a	71–95% +	50–95% +	70–92% +
OKT8, Leu 2a	0–36% +	0–30% +	0–10% +

suggest that normal function of the immune system may also be altered in HTLV-I–infected individuals. Recently a cytotoxic T-cell line derived from an ATL patient was described that was able to specifically kill target cells of the appropriate HLA phenotype and also to express HTLV-I antigens (*21*). However, a cloned cell line possessing integrated HTLV-I provirus derived from these cytotoxic cells was shown to die when exposed to target HTLV-I–infected cells of the appropriate HLA phenotype (*22*). This response is clearly an abberration of normal function and may have implications for the mechanism of HTLV-induced immune suppression.

Additional demonstrations of altered immune function following HTLV-I or -II infection were shown with other T-helper and T-cytotoxic cells. A cloned human helper T-cell line, specific for soluble antigen and certain histocompatible antigen-presenting cells, was observed to lose its specificity following HTLV infection. It subsequently underwent proliferation and also stimulated immunoglobulin production regardless of the presence of soluble antigen or the nature of the antigen-presenting cells (*23*). Similarly, cytotoxic T cells were seen to lose their cytotoxic capability following HTLV infection.

These evidences of altered immune function were acquired in *in vitro* systems. Some suggestions of possible *in vivo* immunosuppressive effects of HTLV-I were reported by Essex and co-workers (*24*). They found that the prevalence of HTLV-I–specific antibodies was significantly greater among Japanese patients hospitalized for common infectious diseases than among random healthy Japanese, even within the HTLV-I endemic area. The implication was that the retrovirus was playing a role in lessening host defenses against various infectious agents. In addition, a number of clinical investigators have noted a high incidence of opportunistic infection in ATL patients (*25, 26*). Together with the data from the cat system showing that the feline leukemia virus causes immunosuppression (*27*), presumably via the p15E protein, these data with HTLV suggested that the virus or another human T-lymphotropic retrovirus might play a role in AIDS. Other reasons for our first proposing this idea are described in the AIDS section.

III. Possible Mechanisms of HTLV-I and HTLV-II–Induced Transformation

The mechanism by which HTLV-I and -II cause transformation *in vitro* and leukemia *in vivo* is not yet clear. Their ability to transform T cells rapidly *in vitro* is a property associated with acute leukemia viruses, which contain cell-derived transforming (or *onc*) genes, and not with chronic leukemia viruses. However, like the chronic leukemia viruses, they do not contain cell-derived *onc* genes (*28*), and leukemic cells and derived cell lines are clonal with respect to proviral integration sites (*29–31*). Despite the clonality of the integration site, HTLV-I does not appear to integrate within a specific locus. No common integration site has been found in any two patients, and in different patients they are even present on different chromosomes (*32*). This differs from the situation in chronic leukemia viruses, where specific proviral integration sites have been reported in tumors. Thus, avian leukosis virus integrates near the *c-myc* locus in bursal lymphomas of chickens (*33–36*) and near *c-erbB* in erythroleukemias (*37*).

DNA sequencing of the proviral DNA has shown, however, that the HTLV-I provirus contains an extra region besides the *gag, pol,* and *env* genes, which was called the *pX* region (*28*) (Fig. 1). This region contained four open reading frames, called pX I–IV. We have shown by DNA sequencing that an HTLV-I variant called HTLV-Ib, isolated from a Zairian patient with ATL (*38*), contains an 11-bp deletion which eliminates the *pX*-I open reading frame (L. Ratner *et al.*, submitted) (Fig. 1). Since HTLV-Ib transforms cells with a normal efficiency, *pX*-I is not essential for transformation.

HTLV-II has been shown to be related over most of the genomes to HTLV-I under nonstringent conditions (*39*). By varying the stringency of hybridization, the most related region was found to be near the 3′ end of the provirus (*40*). DNA sequencing has shown that there is a large open reading frame (*lor*) in the HTLV-II genome (Fig. 2A) which is the most closely related region of HTLV-I and-II (75% nucleic acid homology) (*41*). The *lor* region is in the 3′ 60–70% of the *px*

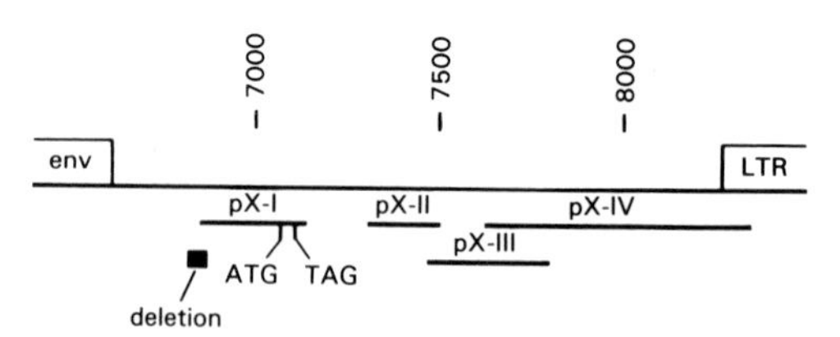

Fig. 1. Deletion in *pX*-I in HTLV-Ib. The relative size and positions of the four open reading frames, each initiated with ATG codons, *pX*-I to *pX*-IV, as suggested by Seiki and coworkers (*28*) are shown schematically. Also indicated is the site of the 11-nucleotide deletion in *pX*-I of HTLV-Ib, which includes the potential initiator ATG codon, together with the position of the next ATG codon in this gene, which is followed after 10 codons by a base pair change predicting a termination codon.

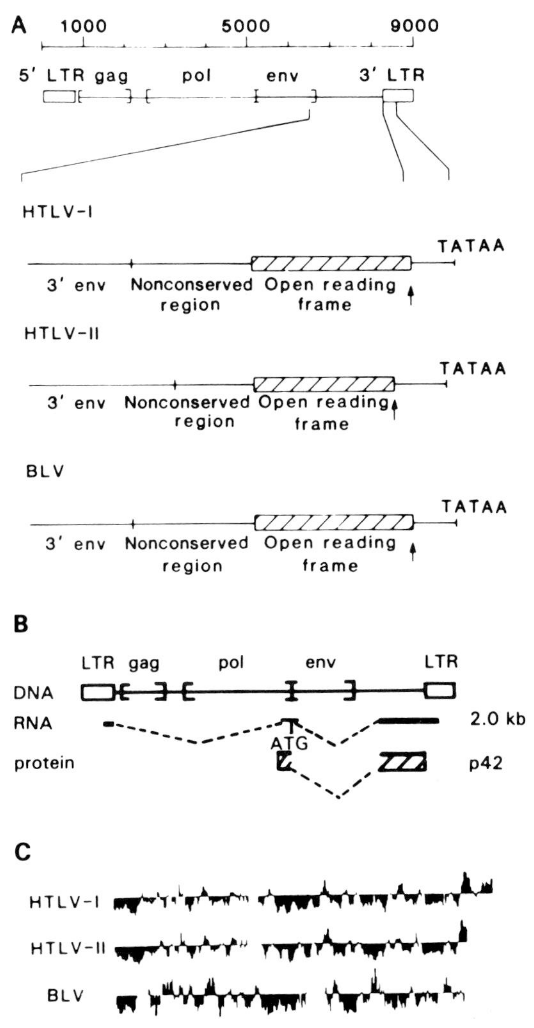

Fig. 2. The open reading frames of the HTLV and BLV genomes. (A) The position of 3′ open reading frames in the genomes of HTLV-I and HTLV-II and of BLV. The 3′ end of the envelope gene is shown, as well as the 5′ terminus of the LTR (↑) and the promoter (TATAA) sequence. The positions of the nonconserved regions and the open reading frames (hatched boxes) are displayed. (B) A model for the splicing scheme of the *lor* mRNA is shown. The 42-kdal protein product encoding this region is assumed to include amino acids encoded by upstream viral sequences. (C) The relative hydophilicity profile of the 3′ open reading frame products of HTLV-I, HTLV-II, and BLV calculated according to the method of Hopp and Woods. Hydrophilic regions are shown above the axis, hydrophobic regions below. Reprinted with modifications from Haseltine *et al.* (*41*).

region. This region is transcribed in the form of a spliced mRNA that includes sequences derived from the *pol* and *env* regions (*42*) (Fig. 2B). The *lor* sequences would predict a protein of at least 38 kdal. A 42-kdal protein has been reported (*43–45*) that is recognized by sera from patients infected with HTLV-I but not by control sera. This is the only viral protein detected in one HTLV-I–transformed nonproducer cell line. Interestingly, bovine leukemia virus (BLV) has recently been shown to have an open reading frame similar in size to *lor* 3′ to the *env* gene (*46*), and the profile of hydrophobic and hydrophilic residues of the inferred amino acid sequences is similar to that of HTLV-I and -II (Fig. 2C).

The precise role of the *pX* or *lor* gene product in the role of the virus is not yet known, but suggestive data have recently been reported from *in vitro* studies with the RNA polymerase promoter contained in the large terminal repeat (LTR) sequences of the viral genome. The gene for the enzyme chloramphenicol acetyl-transerase (CAT) has been put into a plasmid under the transcriptional control of the viral promoter (*47*). The CAT activity after transfection of this plasmid into various cell types is proportional to the rate of CAT mRNA synthesis (*48–50*) and thus indicates the efficiency of the viral promoter. The level of CAT activity or CAT mRNA using the HTLV-I promoter is much higher relative to an SV40 promoter in fibroblasts or lymphoid cells already infected with HTLV-I or lymphoid cells infected with HTLV-II than in other cell types. This is also the case in an HTLV-I nonproducer cell line that expresses only the 42-kdal putative *lor* product (*51, 52*). The HTLV-II promoter is most active in HTLV-II infected cells, much less active in HTLV-I infected cells, and without detectable activity in uninfected cells. These data suggest that the *lor* product may directly or indirectly activate the viral RNA polymerase promoter or enhancer sequences in *trans*.

If the *lor* product can regulate the levels of viral RNA transcription, it is conceivable that it may also act in the same fashion on cellular regulatory elements which control the expression of genes responsible for T-cell function or proliferation. Several cellular gene products are always highly expressed in HTLV-transformed cells. This includes the receptor for the TCGF binding protein (*16, 20*) as well as its mRNA (*53*). Constitutive high levels of expression of this receptor could help drive T-cell proliferation in an unregulated fashion. Another gene (*HT-3*) has been identified as a mRNA that is highly expressed specifically in HTLV-I– and HTLV-II–infected T cells (*54*). Although its function is not yet known, it is tempting to speculate that it is involved somehow in T-cell growth regulation. A third protein present at high levels on HTLV-I infected T cells is HLA-DR (*12, 17*). Figure 3 schematically presents the hypothesis that

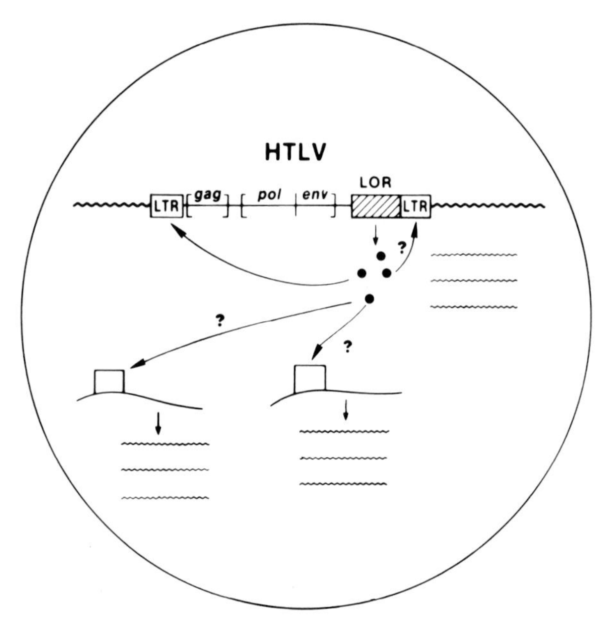

Fig. 3. A model is proposed for transformation mediated by the *trans* activation of the viral LTR and other cellular loci directly or indirectly by the *lor* protein.

the *lor* gene product directly or indirectly activates the transcription of cellular and viral genes in *trans*.

These *in vitro* studies help elucidate the mechanisms of HTLV-induced transformation, but they may not tell the entire story. *In vivo,* a long latent period follows infection with HTLV prior to disease development. In addition, freshly obtained cells of ATL patients often contain chromosome abnormalities, whereas T cells transformed *in vitro* generally have a normal karyotype (*55*). Thus, leukemogenesis may require an additional event. The fact that fresh leukemic cells do not generally express viral mRNA, including that from the *lor* region (*56*), and the lack of a conserved viral integration site (*32*) suggest that the virus may not be necessary for maintenance of the transformed state. Further study will be necessary in order to answer these questions.

IV. HTLV-III: The Cause of AIDS

AIDS was first recognized as a unique clinical syndrome manifested by opportunistic infections or neoplasms complicating an underlying defect in the cellular immune system. The main problem in AIDS is a depletion of the T4 cell population, although the first definition of AIDS was based on the occurrence of opportunistic infections. Most patients present with symptoms resulting from these infections, including fever, pulmonary infiltrates, disseminated cryptococcal disease, and mental status changes associated with central nervous system lesions. However, the common clinical characteristics of patients who develop AIDS are extrainguinal lymphadenopathy and depletion of T cells resulting in clinical manifestations of T-cell deficiency (*57*). Together these common early signs of infection are termed lymphadenopathy syndrome (LAS), AIDS-related complex (ARC), or pre-AIDS.

We believed that AIDS was likely to be caused by a human T-lymphotropic virus for sevral reasons. The epidemiology of AIDS suggested the involvement of an infectious agent, particularly in the cases involved with transfusions. The fact that Factor VIII, which is filtered in a way that should remove bacteria and fungi, was associated with AIDS in hemophiliacs suggested that such an agent would probably be a virus. The specific involvement of OKT4 cells in the disease suggested that an HTLV-like virus was an attractive candidate. This seemed especially so in view of the immunosuppressive effects of HTLV-I infection and opportunistic infections in ATL (see above).

We began to test this hypothesis in 1982, using the same approach that we used for the detection, isolation, and characterization of HTLV-I. Transient reverse transcriptase activity and suggestive electron micrographs were obtained from T cells cultured from blood, lymph nodes, or bone marrow of AIDS or pre-AIDS patients, but we could not maintain the growth of the T cells.

As mentioned above, while HTLV-I and -II cause proliferation of the T cells they infect, HTLV-III causes profound cytopathic effects on T cells, generally leading to cell death. This property made isolation of the virus very difficult due to lack of a system in which to propagate the agent. Eventually a number of neoplastic T-cell lines were tested, and one (HT) was found to support the growth of this virus, which we called HTLV-III. T-cell clones were obtained from the HT cell line, and several of these, including clone H9, were and continue to be better producers of HTLV-III than the parental HT line. The identification of a cell line permissive for HTLV-III infection yet resistant to its cytopathic effects permitted the isolation and detailed characterization of the virus (*7*). Prior to the use of the HT cell line as a target for HTLV-III infection, retroviruses were detected in cells of AIDS patients yet could not be characterized due to insufficient material and rapid loss of the virus-producing cultures.

With achievement of long-term growth of an HTLV-III–infected cell line producing high titers of virus, nucleic acid probes and immunologic reagents could be prepared and used to characterize the agent. Then for the first time seroepidemiological studies could be carried out, virus expression could be assayed in infected cells and tissues, and the molecular biology of the virus became amenable to study.

One of the first findings concerning HTLV-III was the reproducibility with which it could be isolated from cells of patients with AIDS or the AIDS-related complex. Using standard methods of retrovirus detection that had been proven to be successful for detection of HTLV-I and -II, peripheral blood and bone marrow of AIDS patients and those at risk for AIDS were cultured for the purpose of virus isolation. HTLV-III was detected by reverse transcriptase activity, by visualization of virus particles electron microscopically, by virus transmission to permissive cells, and by detection of specific viral antigens using patient sera and, eventually, a rabbit antiserum possessing specific HTLV-III antibodies (*7, 8*). A summary of all the HTLV-III isolates is presented in Table III. Clearly, HTLV-III has been detected in the majority of patients. Using freshly obtained specimens and repeat samples if necessary, viral detection approaches 100% among AIDS patients (S. Salahuddin *et al.*, unpublished observation). This reproducible isolation of HTLV-III from individuals with the disease and with associated risk factors is the first evidence for an etiologic relationship of HTLV-III with AIDS.

The main biological effect of HTLV-III on target cells is primarily a cytopathic one. The main targets of HTLV-III infection are OKT4/Leu 3a-positive T cells, which are also the targets of HTLV-I and -II. Following infection with HTLV-III, a burst of virus production occurs within 1–2 weeks, accompanied by cellular changes including formation of giant multinucleated cells. Subsequently, the OKT4/Leu 3a-positive cells rapidly die. It is likely but not yet certain that other cell types may be infected at low levels with HTLV-III and may serve as *in vivo* reservoirs for continued virus replication and spread.

Additional biological effects occur *in vivo* as a result of direct or indirect manifestations of HTLV-III infection. Before describing these changes, it is first necessary to summarize the evidence for involvement of the virus in AIDS. As mentioned above, the first evidence is the reproducible detection and isolation of HTLV-III from the cells of AIDS patients. Additional strong evidence has been gathered from extensive seroepidemiological studies, summarized in Table IV. In brief, substantial HTLV-III antibody prevalence occurs in groups at risk for

TABLE III

Summary of HTLV-III Isolates

Patients and donors	Diagnosis	Number of HTLV-III Isolates[a]
Homosexual		
Males	AIDS	33
	Pre-AIDS	20
	Clinically normal	13
Nonhomosexual		
Intravenous drug users	Pre-AIDS	2
Hemophiliacs and other transfusion recipients	AIDS	4
	Pre-AIDS	1
Juveniles	AIDS	3
	Pre-AIDS	1
Mothers of juvenile AIDS	AIDS	1
	Pre-AIDS	1
	Clinically normal	2
Promiscuous males	AIDS	2
	Pre-AIDS	6
Spouses of AIDS, pre-AIDS patients	Pre-AIDS	1
	Clinically normal	2
Random donors	Clinically normal	0

[a] The incidence of virus isolation varied depending on condition of cells, health of donor, etc. For AIDS the overall incidence of successful virus isolation/number of patients tested was ~50% and ~80% for pre-AIDS patients.

AIDS, including homosexual males, hemophiliacs, intravenous drug users, and patients with frank AIDS or the AIDS-related complex (*58, 59*). The association of HTLV-III antibodies with AIDS is so strong that a recent study found 100% of all AIDS patients antibody positive compared to 0% of the control subjects and patients with unrelated illnesses (*60*). An etiologic relationship of HTLV-III with AIDS has been further substantiated by serological studies that have suggested the transmission of HTLV-III and subsequent development of the disease. These studies have included recipients of blood transfusions or blood products, a recipient of a kidney transplant from an antibody-positive donor, and heterosexual partners of patients with AIDS. In contrast, numerous healthy controls and patients with other diseases, both malignant and nonmalignant, have been antibody negative. Recent findings that not only is HTLV-III present in the T cells of AIDS patients, explaining the viral transmission by blood or blood products, but

TABLE IV

Antibodies to HTLV-III in Sera of Patients with AIDS and AIDS-Related Complex

Patients/donors	Number positive / number tested (%)
Patients with AIDS	288/297 (97)
Patients with AIDS-related complex	327/360 (91)
Asymptomatic homosexual men	96/235 (41)
High-risk blood donors	9/9
Renal transplant recipient from high-risk donor	1/1
Controls	
Random normal donors	0/238
Black women from Baltimore (1962 collection)	0/100
Healthy Japanese from HTLV-I endemic regions	0/123
Patients with leukemias and other malignancies from HTLV-I endemic areas	0/34
Healthy Surinamese (methadone treatment group)	0/54
Schizophrenics	0/30
Hodgkin's patients and siblings	0/160
Renal transplant recipients	0/24
Other immunosuppressed cases	0/34
Miscellaneous patients[a]	0/82

[a] These include heavily transfused patients and those with hepatitis B virus infection, primary stage syphilis, rheumatoid arthritis, systemic lupus erythematosus, acute mononucleosis, lymphatic leukemias, B- and T-cell lymphomas, alopecia areata, and idiopathic splenomegaly.

also present in saliva (*61*) and in lymphocytes occurring in semen (*62, 63*) explain various transmission routes inferred to occur based on serological data (*64*). The efficiency by which the virus can be transmitted by each of these routes has not yet been elucidated.

While it is clear from *in vitro* studies that HTLV-III kills T cells, recent results using molecularly cloned probes of HTLV-III (*65–67*) indicate that other manifestations of LAS, ARC, or pre-AIDS may be attributable to indirect or secondary effects of virus infection. Thus, when lymph node tissues and peripheral blood lymphocytes of AIDS patients and LAS patients were examined, only a few samples contained detectable levels of HTLV-III provirus, representing at the most 1 infected cell in 100. This same finding was substantiated in *in situ* hybridization experiments, which also indicated that very few lymph node cells were infected with HTLV-III (*67*), or expressing viral RNA (M. Harper *et al.*, submitted). The conclusion of these studies must be that the extensive lymphocyte proliferation is an indirect effect following viral infection. In the same fashion, a direct role of HTLV-III in the induction of Kaposi's sarcoma in AIDS patients was ruled out by the failure to find HTLV-III proviral sequences in the tumor tissue, again suggesting this malignancy must be a secondary indirect effect following viral infection (*67*).

V. Molecular Mechanisms of HTLV-III: Biological Effects

The crucial question facing scientists investigating AIDS is the mechanism by which it causes cell death. With this information in hand, appropriate treatment regimens could be devised to interrupt this effect. While, as discussed above, sequence differences exist between HTLV-III and its related family members and antigenic differences exist between the various protein products of the viral genes, the viruses are remarkably similar; hence they are grouped into a family. The genomic structures of HTLV-I, -II, and -III share some important features in common. Leaving aside for a moment the products of the *gag, pol,* and *env* genes, the prime focus of research interest here, as with the proliferative viruses, is the *pX* region and its protein product. It is certainly possible that some sort of *trans* activation of a cellular gene product could occur in the HTLV-III system, as postulated for HTLV-I and -II. It is also possible that the viral structural proteins themselves may also play a role in causing biological effects. For example, in the feline leukemia virus system, in which retroviral-induced immunosuppression was first recognized, it is the small viral envelope protein, the p15E, that is suspected of causing immunosuppression. Current speculation has also focused on the possibility that the more rapid replication of HTLV-III compared to its sibling viruses may lead to viral strains with increased virulence. Molecular analysis of several HTLV-III isolates has indicated that no two isolates are identical by restriction enzyme analysis (*67*). While the differences so far seem to be minor, it must be determined in which viral genetic regions significant change occurs and whether the observed sequence changes lead to viral protein products with substantial alterations in function or in antigenic properties. This question is particularly important with regard to modulation of the immune response as an effective protective mechanism. It also has serious implications for the production of effective vaccines.

VI. Future Directions

Work to date on HTLV-I, -II, and -III has clearly delineated two areas that future research efforts should take. The first is that of prevention and treatment of the diseases known to be associated with family members. The second, perhaps not so obvious, direction is that of identification of additional family members that may play a role in other human diseases. Considering the latter, some data have already suggested that immunologically related human retroviruses exist, based on weak serological reactivities with known HTLV family members. For example, low-titer antibodies have been detected in patients with early stages of cutaneous T-cell leukemias and lymphomas, such as mycosis fungoides (C. Saxinger *et al.*, unpublished observations). In addition, a high prevalence of low-titer antibody to HTLV-III antigens has been observed in Ugandan children with no evidence of disease, again suggesting that perhaps another cross-reacting virus is the actual infecting agent (C. Saxinger *et al.*, submitted). Alternatively, it may be that the low levels of antibody reflect a continuous exposure to low levels of virus of people in this region, perhaps from a nonhuman viral reservoir. Other human diseases known to occur in clusters, for which no infective agent has yet been found, may also be caused by an as yet unidentified human retrovirus related to the HTLV family. Now that human retroviruses are known to exist, perhaps these other agents can be sought with more confidence. Hodgkin's disease is one, for example, in which a viral etiology would not be unexpected.

Concerning treatment, the proliferative disease caused by HTLV-I (namely ATL), and the cytopathic one caused by HTLV-III, namely AIDS, will clearly require different therapeutic approaches. Studies to date have suggested that HTLV-I, while initiating the transformation of T cells and thereby the process leading to frank malignancy, may not be necessary for the maintenance of the transformed state. Viral mRNA is not detectably expressed in fresh ATL samples, for example (*56*). Thus, treatment regimens should not necessarily be directed toward viral proteins. In the case of AIDS, on the other hand, patients appear to be viremic. The fact that the virus kills T cells indicates that it must continually infect new cells in order to replicate. Therefore, therapeutic regimens aimed at inhibiting HTLV-III replication via its reverse transcriptase may be effective in some stages of disease. Some initial interest in suramin as a reverse transcriptase inhibitor has been expressed and applied to AIDS cells (*68*). It is also clear, however, that as a retrovirus, HTLV-III may exist in the latent proviral form within infected T cells where anti–reverse transcriptase therapy would have no effect. Thus, some means must be found to treat the infected cell as well.

Preventative measures for all HTLV family member infections have focused on vaccines. These approaches require detailed knowledge of the populations at risk of infection, age at first infection, and whether neutralizing antibodies induced by vaccines can be expected to have an effect. Answers to these questions are being sought by extensive serological investigations. If appropriate, it is expected that effective vaccines can be generated using biochemical and molecular biological techniques to identify and synthesize appropriate peptides for generation of an *in vivo* response in humans.

References

1. B. J. Poiesz, F. W. Ruscetti, A. F. Gazdar, P. A. Bunn, J. D. Minna, and R. C. Gallo, *Proc. Natl. Acad. Sci. U.S.A.* **77,** 7415 (1980).
2. M. Robert-Guroff and R. C. Gallo, *Blut* **47,** 1 (1983).
3. B. H. Hahn, G. M. Shaw, M. Popovic, A. Lo Monico, R. C. Gallo, and F. Wong-Staal, *Int. J. Cancer* **34,** 613 (1984).
4. I. Miyoshi, S. Yoshimoto, M. Fujishita, *et al., Lancet* **2,** 658 (1982).
5. H.-G. Guo, F. Wong-Staal, and R. C. Gallo, *Science* **223,** 1195 (1984).
6. V. S. Kalyanaraman, M. G. Sarngadharan, M. Robert-Guroff, *et al., Science* **218,** 571 (1982).
7. M. Popovic, M. G. Sarngadharan, E. Read, and R. C. Gallo, *Science* **224,** 497 (1984).
8. R. C. Gallo, S. Z. Salahuddin, M. Popovic, *et al., Science* **224,** 500 (1984).
9. R. C. Gallo, *in* "Cancer Surveys" (L. M. Franks, L. M. Wyke, and R. C. Weiss, eds.) Vol. 3., p. 113. Oxford Univ. Press, London, 1984.
10. I. Miyoshi, I. Kubonishi, S. Yoshimoto, *et al., Nature (London)* **294,** 770 (1981).
11. P. D. Markham, S. Z. Salahuddin, V. S. Kalyanaraman, M. Popovic, P. Sarin, and R. C. Gallo, *Int. J. Cancer* **31,** 413 (1983).
12. M. Popovic, P. S. Sarin, M. Robert-Guroff, *et al., Science* **219,** 856 (1983).
13. M. Popovic, G. Lange-Wantzin, P. S. Sarin, D. Mann, and R. C. Gallo, *Proc. Natl. Acad. Sci. U.S.A.* **80,** 5402 (1983).
14. S. Z. Salahuddin, P. D. Markham, F. Wong-Staal, G. Franchini, V. S. Kalyanaraman, and R. C. Gallo, *Virology* **129,** 51 (1983).
15. F. W. Ruscetti, M. Robert-Guroff, L. Ceccherini-Nelli, J. Minowada, M. Popovic, and R. C. Gallo, *Int. J. Cancer* **31,** 171 (1983).
16. W. C. Greene and R. J. Robb, *in* "Contemporary Topics in Molecular Immunology" (S. Gillis, ed.), Plenum, New York, in press.

17. D. L. Mann, M. Popovic, P. Sarin, *et al., Nature (London)* **305,** 58 (1983).
18. P. D. Markham, S. Z. Salahuddin, B. Macchi, M. Robert-Guroff, and R. C. Gallo, *Int. J. Cancer* **33,** 13 (1984).
19. S. Z. Salahuddin, P. D. Markham, S. G. Lindner, *et al., Science* **223,** 703 (1984).
20. J. E. Gootenberg, F. W. Ruscetti, J. W. Mier, A. F. Gazdar, and R. C. Gallo, *J. Exp. Med.* **154,** 1403 (1981).
21. H. Mitsuya, L. A. Matis, M. Megson, *et al., J. Exp. Med.* **158,** 994 (1983).
22. H. Mitsuya, H.-G. Guo, M. Megson, C. Trainor, M. S. Reitz, and S. Broder, *Science* **223,** 1293 (1984).
23. M. Popovic, N. Flomenberg, D. J. Volkman, *et al., Science* **226,** 459 (1984).
24. M. Essex, M. F. McLane, N. Tachibana, D. P. Francis, and T. H. Lee, *in* "Human T-Cell Leukemia/Lymphoma Virus" (R. C. Gallo, M. Essex, and L. Gross, eds.), p. 355. Cold Spring Harbor Laboratory, New York, 1984.
25. K. Kinoshita, S. Kamihara, Y. Yamada, *et al., in* "Adult T-Cell Leukemia and Related Diseases" (M. Hanaoka, K. Takatsuki, and M. Shimoyama, eds.), p. 167. Plenum, Tokyo, 1982.
26. N. Ueda, K. Iwata, H. Tokuoka, T. Akagi, J. Ito, and M. Mizushima, *Acta Pathol. Jpn.* **29**(2), 221 (1979).
27. Z. Trainin, D. Wernicke, H. Ungar-Waron, and M. Essex, *Science* **220,** 858 (1983).
28. M. Seiki, S. Hattori, Y. Hirayama, and M. Yoshida, *Proc. Natl. Acad. Sci. U.S.A.* **80,** 3618 (1983).
29. M. Yoshida, I. Miyoshi, and Y. Hinuma, *Proc. Natl. Acad. Sci. U.S.A.* **79,** 2031 (1982).
30. F. Wong-Staal, B. Hahn, V. Manzari, *et al., Nature (London)* **302,** 626 (1983).
31. M. Yoshida, M. Seiki, K. Yamaguchi, and K. Takatsuki, *Proc. Natl. Acad. Sci. U.S.A.* **81,** 2534 (1984).
32. M. Seiki, R. Eddy, T. R. Shows, and H. Yoshida, *Nature (London)* **309,** 640 (1984).
33. Y. K. Fung, A. N. Fadley, L. B. Crittenden, and H. J. Kung, *Proc. Natl. Acad. Sci. U.S.A.* **78,** 3418 (1981).
34. W. S. Hayward, B. G. Neel, and S. M. Astrin, *Nature (London)* **290,** 475 (1981).
35. B. G. Neel, W. S. Hayward, H. L. Robinson, J. Fang, and S. M. Astrin, *Cell* **23,** 323 (1981).
36. G. S. Payne, J. M. Bishop, and M. E. Varmus, *Nature (London)* **295,** 209 (1982).
37. Y. K. T. Fung, W. G. Lewis, L. B. Crittenden, and J. H. Kung, *Cell* **33,** 357 (1983).
38. B. H. Hahn, G. M. Shaw, M. Popovic, *et al., Int. J. Cancer* in press.
39. E. P. Gelmann, G. Franchini, V. Manzari, F. Wong-Staal, and R. C. Gallo, *Proc. Natl. Acad. Sci. U.S.A.* **81,** 993 (1984).
40. G. M. Shaw, M. A. Gonda, G. H. Flickinger, B. H. Hahn, R. C. Gallo, and F. Wong-Staal, *Proc. Natl. Acad. Sci. U.S.A.* **81,** 4544 (1984).
41. W. A. Haseltine, J. Sodroski, R. Patarca, D. Briggs, D. Perkins, and F. Wong-Staal, *Science* **225,** 419 (1984).
42. W. Wachsman, K. Shimotoho, S. C. Clark, *et al., Science* **226,** 177 (1984).
43. M. Miwa, K. Shimotohno, H. Hoshino, M. Fujino, and T. Sugimura, *Gann* **75,** 752 (1984).
44. T. H. Lee, J. E. Coligan, J. G. Sodroski, *et al., Science* **226,** 57 (1984).
45. D. J. Slamon, K. Shimotohno, M. J. Cline, D. W. Golde, and I. S. Y. Chen, *Science* **226,** 61 (1984).
46. N. R. Rice, R. M. Stephenson, D. Couez, *et al., Virology* **138,** 82 (1984).
47. J. G. Sodroski, C. A. Rosen, and W. A. Haseltine, *Science* **225,** 381 (1984).
48. C. M. Gorman, L. F. Moffat, and B. H. Howard, *Mol. Cell. Biol.* **2,** 1044 (1982).

49. C. M. Gorman, T. Merlino, M. C. Willingham, I. Pastan, and B. Howard, *Proc. Natl. Acad. Sci. U.S.A.* **79,** 6777 (1982).
50. M. D. Walker, R. Edlund, A. M. Boulet, and W. J. Rutter, *Nature (London)* **306,** 557 (1983).
51. T. H. Lee, J. E. Coligan, T. Homma, *et al., Science* **220,** 57 (1984).
52. S. Z. Salahuddin, P. D. Markham, F. Wong-Staal, G. Franchini, V. S. Kalyanaraman, and R. C. Gallo, *Virology* **129,** 51 (1983).
53. W. J. Leonard, J. M. Depper, G. R. Crabtree, *et al., Nature (London)* **311,** 626 (1984).
54. V. Manzari, R. C. Gallo, G. Franchini, *et al., Proc. Natl. Acad. Sci. U.S.A.* **79,** 2490 (1982).
55. P. C. Nowell, J. B. Finan, J. Clark, Jr., P. S. Sarin, and R. C. Gallo, *J. Natl. Cancer Inst.* in press.
56. G. Franchini, F. Wong-Staal, and R. C. Gallo, *Proc. Natl. Acad. Sci. U.S.A.* **81,** 6207 (1984).
57. A. S. Fauci, A. M. Macher, D. L. Longo, *et al., Ann. Intern. Med.* **100,** 92 (1984).
58. J. Schupbach, M. Popovic, R. V. Gilden, M. A. Gonda, M. G. Sarngadharan, and R. C. Gallo, *Science* **224,** 503 (1984).
59. M. G. Sarngadharan, M. Popovic, L. Bruch, J. Schupbach, and R. C. Gallo, *Science* **224,** 506 (1984).
60. B. Safai, M. G. Sarngadharan, J. E. Groopman, *et al., Lancet* **1,** 1438 (1984).
61. J. E. Groopman, S. Z. Salahuddin, M. G. Sarngadharan, *et al., Science* **226,** 447 (1984).
62. D. Zagury, J. Bernard, J. Leibowitch, *et al., Science* **226,** 449 (1984).
63. D. D. Ho, R. T. Schooley, T. R. Rota, *et al., Science* **226,** 451 (1984).
64. J. J. Goedert, M. G. Sarngadharan, R. J. Biggar, *et al., Lancet* **2,** 711 (1984).
65. S. K. Arya, R. C. Gallo, B. Hahn, *et al., Science* **225,** 927 (1984).
66. B. H. Hahn, G. M. Shaw, S. K. Arya, M. Popovic, R. C. Gallo, and F. Wong-Staal, *Nature (London)* in press.
67. G. M. Shaw, B. H. Hahn, S. K. Arya, J. E. Groopman, R. C. Gallo, and F. Wong-Staal, *Science* **226,** 1165 (1984).
68. H. Mitsuya, M. Popovic, R. Yarchoan, S. Matsushita, R. C. Gallo, and S. Broder, *Science* **226,** 172 (1984).

Index